Calculus

Supplement

James Stewart
McMaster University

BROOKS/COLE
THOMSON LEARNING

Printed in the United States of America

Brooks.Cole
511 Forest Lodge Road
Pacific Grove, CA 93950
USA

For information about our products, contact us:
Thomson Learning Academic Resource Center
1-800-423-0563
http://www.brookscole.com

International Headquarters
Thomson Learning
International Division
290 Harbor Drive, 2nd Floor
Stamford, CT 06902-7477
USA

UK/Europe/Middle East/South Africa
Thomson Learning
Berkshire House
168-173 High Holborn
London WCIV 7AA

Asia
Thomson Learning
60 Albert Street, #15-01
Albert Complex
Singapore 189969

Canada
Nelson Thomson Learning
1120 Birchmount Road
Toronto, Ontario MIK 5G4
Canada
United Kingdom

0-534-48102-7

The Adaptable Courseware Program consists of products and additions to existing Brooks/Cole products that are produced from camera-ready copy. Peer review, class testing, and accuracy are primarily the responsibility of the author(s).

7.7 Second-Order Linear Equations

A **second-order linear differential equation** has the form

$$\boxed{1} \qquad P(x)\frac{d^2y}{dx^2} + Q(x)\frac{dy}{dx} + R(x)y = G(x)$$

where P, Q, R, and G are continuous functions. We saw in Section 10.1 that equations of this type arise in the study of the motion of a spring. In Section 18.3 we will further pursue this application as well as the application to electric circuits.

In this section we study the case where $G(x) = 0$, for all x, in Equation 1. Such equations are called **homogeneous** linear equations. Thus, the form of a second-order linear homogeneous differential equation is

$$\boxed{2} \qquad P(x)\frac{d^2y}{dx^2} + Q(x)\frac{dy}{dx} + R(x)y = 0$$

If $G(x) \neq 0$ for some x, Equation 1 is **nonhomogeneous** and is discussed in Section 18.2.

Two basic facts enable us to solve homogeneous linear equations. The first of these says that if we know two solutions y_1 and y_2 of such an equation, then the **linear combination** $y = c_1y_1 + c_2y_2$ is also a solution.

3 Theorem If $y_1(x)$ and $y_2(x)$ are both solutions of the linear homogeneous equation (2) and c_1 and c_2 are any constants, then the function

$$y(x) = c_1y_1(x) + c_2y_2(x)$$

is also a solution of Equation 2.

Proof Since y_1 and y_2 are solutions of Equation 2, we have

$$P(x)y_1'' + Q(x)y_1' + R(x)y_1 = 0$$

and

$$P(x)y_2'' + Q(x)y_2' + R(x)y_2 = 0$$

Therefore, using the basic rules for differentiation, we have

$$\begin{aligned}
P(x)y'' + Q(x)y' + R(x)y \\
&= P(x)(c_1y_1 + c_2y_2)'' + Q(x)(c_1y_1 + c_2y_2)' + R(x)(c_1y_1 + c_2y_2) \\
&= P(x)(c_1y_1'' + c_2y_2'') + Q(x)(c_1y_1' + c_2y_2') + R(x)(c_1y_1 + c_2y_2) \\
&= c_1[P(x)y_1'' + Q(x)y_1' + R(x)y_1] + c_2[P(x)y_2'' + Q(x)y_2' + R(x)y_2] \\
&= c_1(0) + c_2(0) = 0
\end{aligned}$$

Thus, $y = c_1y_1 + c_2y_2$ is a solution of Equation 2. ■

The other fact we need is given by the following theorem, which is proved in more advanced courses. It says that the general solution is a linear combination of two **linearly independent** solutions y_1 and y_2. This means that neither y_1 nor y_2 is a constant multiple of the other. For instance, the functions $f(x) = x^2$ and $g(x) = 5x^2$ are linearly dependent, but $f(x) = e^x$ and $g(x) = xe^x$ are linearly independent.

4 Theorem If y_1 and y_2 are linearly independent solutions of Equation 2, then the general solution is given by

$$y(x) = c_1 y_1(x) + c_2 y_2(x)$$

where c_1 and c_2 are arbitrary constants.

Theorem 4 is very useful because it says that if we know *two* particular linearly independent solutions, then we know *every* solution.

In general, it is not easy to discover particular solutions to a second-order linear equation. But it is always possible to do so if the coefficient functions P, Q, and R are constant functions, that is, if the differential equation has the form

5 $$ay'' + by' + cy = 0$$

where a, b, and c are constants and $a \neq 0$.

It's not hard to think of some likely candidates for particular solutions of Equation 5 if we state the equation verbally. We are looking for a function y such that a constant times its second derivative y'' plus another constant times y' plus a third constant times y is equal to 0. We know that the exponential function $y = e^{rx}$ (where r is a constant) has the property that its derivative is a constant multiple of itself: $y' = re^{rx}$. Furthermore, $y'' = r^2e^{rx}$. If we substitute these expressions into Equation 5, we see that $y = e^{rx}$ is a solution if

$$ar^2e^{rx} + bre^{rx} + ce^{rx} = 0$$

or

$$(ar^2 + br + c)e^{rx} = 0$$

But e^{rx} is never 0. Thus, $y = e^{rx}$ is a solution of Equation 5 if r is a root of the equation

6 $$ar^2 + br + c = 0$$

Equation 6 is called the **auxiliary equation** (or **characteristic equation**) of the differential equation $ay'' + by' + cy = 0$. Notice that it is an algebraic equation that is obtained from the differential equation by replacing y'' by r^2, y' by r, and y by 1.

Sometimes the roots r_1 and r_2 of the auxiliary equation can be found by factoring. In other cases they are found by using the quadratic formula:

7 $$r_1 = \frac{-b + \sqrt{b^2 - 4ac}}{2a} \qquad r_2 = \frac{-b - \sqrt{b^2 - 4ac}}{2a}$$

We distinguish three cases according to the sign of the discriminant $b^2 - 4ac$.

CASE I ■ $b^2 - 4ac > 0$

In this case the roots r_1 and r_2 of the auxiliary equation are real and distinct, so $y_1 = e^{r_1 x}$ and $y_2 = e^{r_2 x}$ are two linearly independent solutions of Equation 5. (Note that $e^{r_2 x}$ is not a constant multiple of $e^{r_1 x}$.) Therefore, by Theorem 4, we have the following fact.

8 If the roots r_1 and r_2 of the auxiliary equation $ar^2 + br + c = 0$ are real and unequal, then the general solution of $ay'' + by' + cy = 0$ is

$$y = c_1 e^{r_1 x} + c_2 e^{r_2 x}$$

■ In Figure 1 the graphs of the basic solutions $f(x) = e^{2x}$ and $g(x) = e^{-3x}$ of the differential equation in Example 1 are shown in black and red, respectively. Some of the other solutions, linear combinations of f and g, are shown in blue.

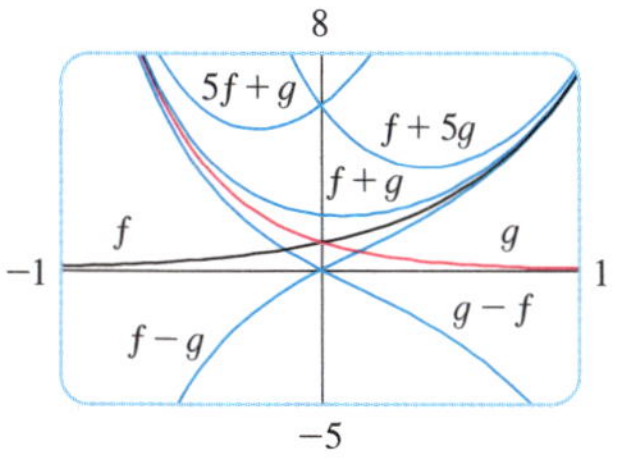

FIGURE 1

EXAMPLE 1 ■ Solve the equation $y'' + y' - 6y = 0$.

SOLUTION The auxiliary equation is

$$r^2 + r - 6 = (r - 2)(r + 3) = 0$$

whose roots are $r = 2, -3$. Therefore, by (8) the general solution of the given differential equation is

$$y = c_1 e^{2x} + c_2 e^{-3x}$$

We could verify that this is indeed a solution by differentiating and substituting into the differential equation. ■

EXAMPLE 2 ■ Solve $3\dfrac{d^2y}{dx^2} + \dfrac{dy}{dx} - y = 0$.

SOLUTION To solve the auxiliary equation $3r^2 + r - 1 = 0$ we use the quadratic formula:

$$r = \frac{-1 \pm \sqrt{13}}{6}$$

Since the roots are real and distinct, the general solution is

$$y = c_1 e^{(-1+\sqrt{13})x/6} + c_2 e^{(-1-\sqrt{13})x/6}$$ ■

CASE II ■ $b^2 - 4ac = 0$

In this case $r_1 = r_2$; that is, the roots of the auxiliary equation are real and equal. Let's denote by r the common value of r_1 and r_2. Then, from Equations 7, we have

9 $$r = -\frac{b}{2a} \qquad \text{so} \qquad 2ar + b = 0$$

We know that $y_1 = e^{rx}$ is one solution of Equation 5. We now verify that $y_2 = xe^{rx}$ is also a solution:

$$\begin{aligned} ay_2'' + by_2' + cy_2 &= a(2re^{rx} + r^2xe^{rx}) + b(e^{rx} + rxe^{rx}) + cxe^{rx} \\ &= (2ar + b)e^{rx} + (ar^2 + br + c)xe^{rx} \\ &= 0(e^{rx}) + 0(xe^{rx}) = 0 \end{aligned}$$

The first term is 0 by Equations 9; the second term is 0 because r is a root of the auxiliary equation. Since $y_1 = e^{rx}$ and $y_2 = xe^{rx}$ are linearly independent solutions, Theorem 4 provides us with the general solution.

10 If the auxiliary equation $ar^2 + br + c = 0$ has only one real root r, then the general solution of $ay'' + by' + cy = 0$ is

$$y = c_1 e^{rx} + c_2 x e^{rx}$$

■ Figure 2 shows the basic solutions $f(x) = e^{-3x/2}$ and $g(x) = xe^{-3x/2}$ in Example 3 and some other members of the family of solutions. Notice that all of them approach 0 as $x \to \infty$.

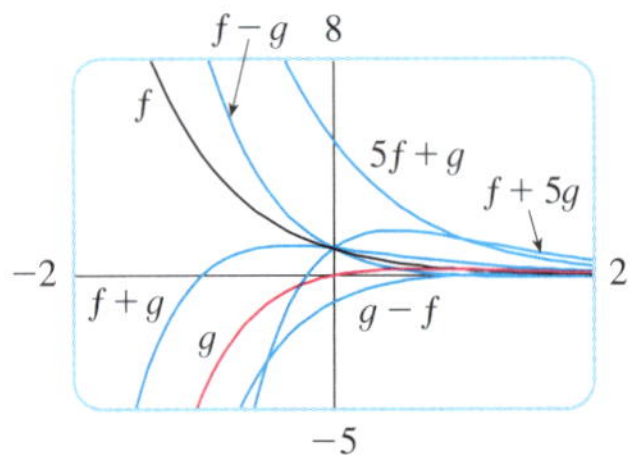

FIGURE 2

EXAMPLE 3 ■ Solve the equation $4y'' + 12y' + 9y = 0$.

SOLUTION The auxiliary equation $4r^2 + 12r + 9 = 0$ can be factored as

$$(2r + 3)^2 = 0$$

so the only root is $r = -\frac{3}{2}$. By (10) the general solution is

$$y = c_1 e^{-3x/2} + c_2 x e^{-3x/2}$$ ■

CASE III ■ $b^2 - 4ac < 0$

In this case the roots r_1 and r_2 of the auxiliary equation are complex numbers. (See Appendix G for information about complex numbers.) We can write

$$r_1 = \alpha + i\beta \qquad r_2 = \alpha - i\beta$$

where α and β are real numbers. [In fact, $\alpha = -b/(2a)$, $\beta = \sqrt{4ac - b^2}/(2a)$.] Then, using Euler's equation

$$e^{i\theta} = \cos\theta + i\sin\theta$$

from Appendix G, we write the solution of the differential equation as

$$\begin{aligned} y &= C_1 e^{r_1 x} + C_2 e^{r_2 x} = C_1 e^{(\alpha + i\beta)x} + C_2 e^{(\alpha - i\beta)x} \\ &= C_1 e^{\alpha x}(\cos\beta x + i\sin\beta x) + C_2 e^{\alpha x}(\cos\beta x - i\sin\beta x) \\ &= e^{\alpha x}[(C_1 + C_2)\cos\beta x + i(C_1 - C_2)\sin\beta x] \\ &= e^{\alpha x}(c_1\cos\beta x + c_2\sin\beta x) \end{aligned}$$

where $c_1 = C_1 + C_2$, $c_2 = i(C_1 - C_2)$. This gives all solutions (real or complex) of the differential equation. The solutions are real when the constants c_1 and c_2 are real. We summarize the discussion as follows.

11 If the roots of the auxiliary equation $ar^2 + br + c = 0$ are the complex numbers $r_1 = \alpha + i\beta$, $r_2 = \alpha - i\beta$, then the general solution of $ay'' + by' + cy = 0$ is

$$y = e^{\alpha x}(c_1\cos\beta x + c_2\sin\beta x)$$

■ Figure 3 shows the graphs of the solutions in Example 4, $f(x) = e^{3x} \cos 2x$ and $g(x) = e^{3x} \sin 2x$, together with some linear combinations. All solutions approach 0 as $x \to -\infty$.

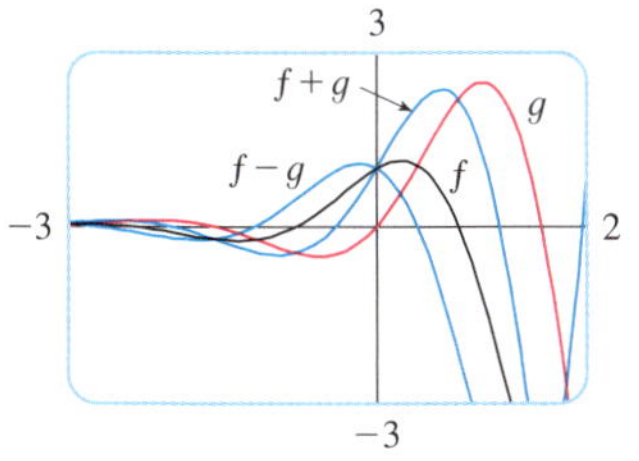

FIGURE 3

EXAMPLE 4 ■ Solve the equation $y'' - 6y' + 13y = 0$.

SOLUTION The auxiliary equation is $r^2 - 6r + 13 = 0$. By the quadratic formula, the roots are

$$r = \frac{6 \pm \sqrt{36 - 52}}{2} = \frac{6 \pm \sqrt{-16}}{2} = 3 \pm 2i$$

By (11) the general solution of the differential equation is

$$y = e^{3x}(c_1 \cos 2x + c_2 \sin 2x)$$ ■

Initial-Value and Boundary-Value Problems

An **initial-value problem** for the second-order Equation 1 or 2 consists of finding a solution y of the differential equation that also satisfies initial conditions of the form

$$y(x_0) = y_0 \qquad y'(x_0) = y_1$$

where y_0 and y_1 are given constants. If P, Q, R, and G are continuous on an interval and $P(x) \neq 0$ there, then a theorem found in more advanced books guarantees the existence and uniqueness of a solution to this initial-value problem. Examples 5 and 6 illustrate the technique for solving such a problem.

A **boundary-value problem** for Equation 1 consists of finding a solution y of the differential equation that also satisfies boundary conditions of the form

$$y(x_0) = y_0 \qquad y(x_1) = y_1$$

In contrast with the situation for initial-value problems, a boundary-value problem does not always have a solution. The method is illustrated in Example 7.

EXAMPLE 5 ■ Solve the initial-value problem

$$y'' + y' - 6y = 0 \qquad y(0) = 1 \qquad y'(0) = 0$$

SOLUTION From Example 1 we know that the general solution of the differential equation is

$$y(x) = c_1 e^{2x} + c_2 e^{-3x}$$

Differentiating this solution, we get

$$y'(x) = 2c_1 e^{2x} - 3c_2 e^{-3x}$$

■ Figure 4 shows the graph of the solution of the initial-value problem in Example 5. Compare with Figure 1.

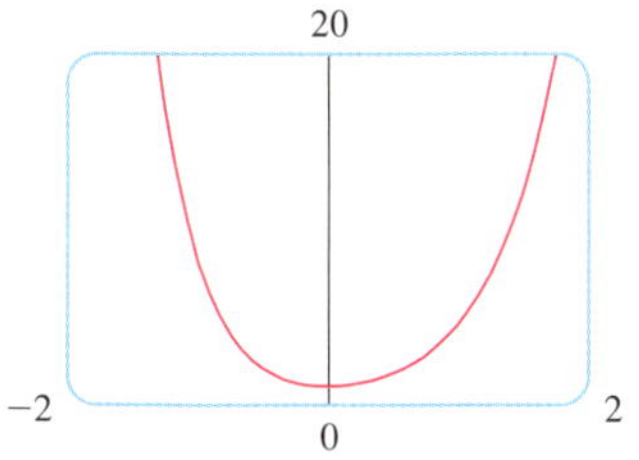

FIGURE 4

To satisfy the initial conditions we require that

12 $$y(0) = c_1 + c_2 = 1$$

13 $$y'(0) = 2c_1 - 3c_2 = 0$$

From (13) we have $c_2 = \frac{2}{3}c_1$ and so (12) gives

$$c_1 + \tfrac{2}{3}c_1 = 1 \qquad c_1 = \tfrac{3}{5} \qquad c_2 = \tfrac{2}{5}$$

Thus, the required solution of the initial-value problem is

$$y = \tfrac{3}{5}e^{2x} + \tfrac{2}{5}e^{-3x}$$ ■

EXAMPLE 6 ■ Solve the initial-value problem

$$y'' + y = 0 \qquad y(0) = 2 \qquad y'(0) = 3$$

SOLUTION The auxiliary equation is $r^2 + 1 = 0$, or $r^2 = -1$, whose roots are $\pm i$. Thus $\alpha = 0$, $\beta = 1$, and since $e^{0x} = 1$, the general solution is

$$y(x) = c_1 \cos x + c_2 \sin x$$

Since $$y'(x) = -c_1 \sin x + c_2 \cos x$$

the initial conditions become

$$y(0) = c_1 = 2 \qquad y'(0) = c_2 = 3$$

Therefore, the solution of the initial-value problem is

$$y(x) = 2 \cos x + 3 \sin x$$ ■

■ The solution to Example 6 is graphed in Figure 5. It appears to be a shifted sine curve and, indeed, you can verify that another way of writing the solution is

$$y = \sqrt{13} \sin(x + \phi) \quad \text{where } \tan \phi = \tfrac{2}{3}$$

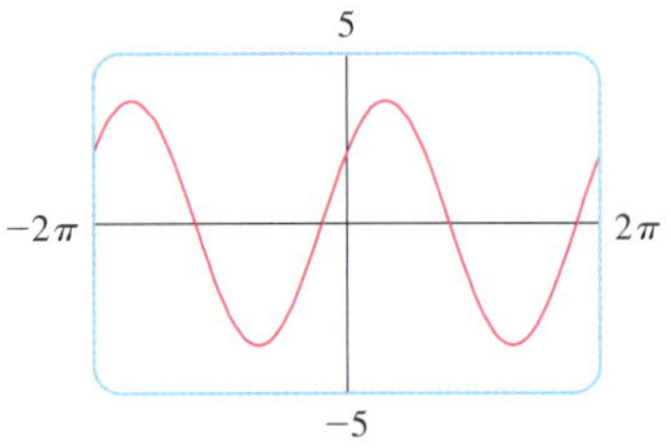

FIGURE 5

EXAMPLE 7 ■ Solve the boundary-value problem

$$y'' + 2y' + y = 0 \qquad y(0) = 1 \qquad y(1) = 3$$

SOLUTION The auxiliary equation is

$$r^2 + 2r + 1 = 0 \qquad \text{or} \qquad (r + 1)^2 = 0$$

whose only root is $r = -1$. Therefore, the general solution is

$$y(x) = c_1 e^{-x} + c_2 x e^{-x}$$

The boundary conditions are satisfied if

$$y(0) = c_1 = 1$$

$$y(1) = c_1 e^{-1} + c_2 e^{-1} = 3$$

The first condition gives $c_1 = 1$, so the second condition becomes

$$e^{-1} + c_2 e^{-1} = 3$$

Solving this equation for c_2 by first multiplying through by e, we get

$$1 + c_2 = 3e \qquad \text{so} \qquad c_2 = 3e - 1$$

Thus, the solution of the boundary-value problem is

$$y = e^{-x} + (3e - 1)xe^{-x}$$ ■

■ Figure 6 shows the graph of the solution of the boundary-value problem in Example 7.

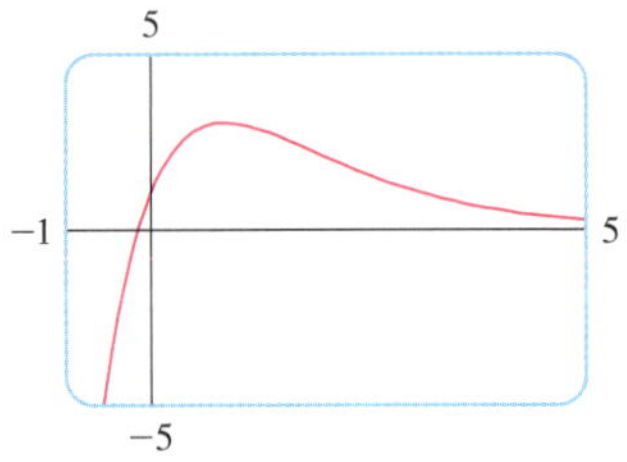

FIGURE 6

Summary: Solutions of $ay'' + by' + c = 0$

Roots of $ar^2 + br + c = 0$	General solution
r_1, r_2 real and distinct	$y = c_1 e^{r_1 x} + c_2 e^{r_2 x}$
$r_1 = r_2 = r$	$y = c_1 e^{rx} + c_2 x e^{rx}$
r_1, r_2 complex: $\alpha \pm i\beta$	$y = e^{\alpha x}(c_1 \cos \beta x + c_2 \sin \beta x)$

7.7 Exercises

1–13 ■ Solve the differential equation.

1. $y'' - 6y' + 8y = 0$

2. $y'' - 4y' + 8y = 0$

3. $y'' + 8y' + 41y = 0$

4. $2y'' - y' - y = 0$

5. $y'' - 2y' + y = 0$

6. $3y'' = 5y'$

7. $4y'' + y = 0$

8. $16y'' + 24y' + 9y = 0$

9. $4y'' + y' = 0$

10. $9y'' + 4y = 0$

11. $\dfrac{d^2y}{dt^2} - 2\dfrac{dy}{dt} - y = 0$

12. $\dfrac{d^2y}{dt^2} - 6\dfrac{dy}{dt} + 4y = 0$

13. $\dfrac{d^2y}{dt^2} + \dfrac{dy}{dt} + y = 0$

14–16 ■ Graph the two basic solutions of the differential equation and several other solutions. What features do the solutions have in common?

14. $6\dfrac{d^2y}{dx^2} - \dfrac{dy}{dx} - 2y = 0$

15. $\dfrac{d^2y}{dx^2} - 8\dfrac{dy}{dx} + 16y = 0$

16. $\dfrac{d^2y}{dx^2} - 2\dfrac{dy}{dx} + 5y = 0$

17–24 ■ Solve the initial-value problem.

17. $2y'' + 5y' + 3y = 0, \quad y(0) = 3, \quad y'(0) = -4$

18. $y'' - 4y = 0, \quad y(0) = 1, \quad y'(0) = 0$

19. $y'' - 2y' + 2y = 0, \quad y(0) = 1, \quad y'(0) = 2$

20. $y'' + 4y' + 6y = 0, \quad y(0) = 2, \quad y'(0) = 4$

21. $y'' - 2y' - 3y = 0, \quad y(1) = 3, \quad y'(1) = 1$

22. $y'' - 2y' + y = 0, \quad y(2) = 0, \quad y'(2) = 1$

23. $y'' + 9y = 0, \quad y(\pi/3) = 0, \quad y'(\pi/3) = 1$

24. $y'' + 4y = 0, \quad y(\pi/6) = 1, \quad y'(\pi/6) = 0$

25–32 ■ Solve the boundary-value problem, if possible.

25. $y'' + 4y' + 4y = 0, \quad y(0) = 0, \quad y(1) = 3$

26. $y'' + 5y' - 6y = 0, \quad y(0) = 0, \quad y(2) = 1$

27. $y'' + y = 0, \quad y(0) = 1, \quad y(\pi) = 0$

28. $y'' + 9y = 0, \quad y(0) = 1, \quad y(\pi/2) = 0$

29. $y'' - y' - 2y = 0, \quad y(-1) = 1, \quad y(1) = 0$

30. $y'' + 4y' + 3y = 0, \quad y(1) = 0, \quad y(3) = 2$

31. $y'' + 4y' + 13y = 0, \quad y(0) = 2, \quad y(\pi/2) = 1$

32. $y'' + 2y' + 5y = 0, \quad y(0) = 1, \quad y(\pi) = 2$

33. (a) Show that the boundary-value problem $y'' + \lambda y = 0$, $y(0) = 0$, $y(L) = 0$ has only the trivial solution $y = 0$ for the cases $\lambda = 0$ and $\lambda < 0$.

(b) For the case $\lambda > 0$, find the values of λ for which this problem has a nontrivial solution and give the corresponding solution.

34. If a, b, and c are all positive constants and $y(x)$ is a solution of the differential equation $ay'' + by' + cy = 0$, show that $\lim_{x \to \infty} y(x) = 0$.

7.8 Nonhomogeneous Linear Equations

In this section we learn how to solve second-order nonhomogeneous linear differential equations with constant coefficients, that is, equations of the form

1 $$ay'' + by' + cy = G(x)$$

where a, b, and c are constants and G is a continuous function. The related homogeneous equation

2 $$ay'' + by' + cy = 0$$

is called the **complementary equation** and plays an important role in the solution of the original nonhomogeneous equation (1).

3 Theorem The general solution of the nonhomogeneous differential equation (1) can be written as

$$y(x) = y_p(x) + y_c(x)$$

where y_p is a particular solution of Equation 1 and y_c is the general solution of the complementary Equation 2.

Proof All we have to do is verify that if y is any solution of Equation 1, then $y - y_p$ is a solution of the complementary Equation 2. Indeed

$$\begin{aligned} a(y - y_p)'' + b(y - y_p)' + c(y - y_p) &= ay'' - ay_p'' + by' - by_p' + cy - cy_p \\ &= (ay'' + by' + cy) - (ay_p'' + by_p' + cy_p) \\ &= g(x) - g(x) = 0 \end{aligned}$$ ■

We know from Section 18.1 how to solve the complementary equation. (Recall that the solution is $y_c = c_1y_1 + c_2y_2$, where y_1 and y_2 are linearly independent solutions of Equation 2.) Therefore, Theorem 3 says that we know the general solution of the nonhomogeneous equation as soon as we know a particular solution y_p. There are two methods for finding a particular solution: The method of undetermined coefficients is straightforward but works only for a restricted class of functions G. The method of variation of parameters works for every function G but is usually more difficult to apply in practice.

The Method of Undetermined Coefficients

We first illustrate the method of undetermined coefficients for the equation

$$ay'' + by' + cy = G(x)$$

where $G(x)$ is a polynomial. It is reasonable to guess that there is a particular solution y_p that is a polynomial of the same degree as G because if y is a polynomial, then $ay'' + by' + cy$ is also a polynomial. We therefore substitute $y_p(x) =$ a polynomial (of the same degree as G) into the differential equation and determine the coefficients.

EXAMPLE 1 ■ Solve the equation $y'' + y' - 2y = x^2$.

SOLUTION The auxiliary equation of $y'' + y' - 2y = 0$ is

$$r^2 + r - 2 = (r - 1)(r + 2) = 0$$

with roots $r = 1, -2$. So the solution of the complementary equation is

$$y_c = c_1e^x + c_2e^{-2x}$$

Since $G(x) = x^2$ is a polynomial of degree 2, we seek a particular solution of the form

$$y_p(x) = Ax^2 + Bx + C$$

Then $y_p' = 2Ax + B$ and $y_p'' = 2A$ so, substituting into the given differential equation, we have

$$(2A) + (2Ax + B) - 2(Ax^2 + Bx + C) = x^2$$

or

$$-2Ax^2 + (2A - 2B)x + (2A + B - 2C) = x^2$$

■ Figure 1 shows four solutions of the differential equation in Example 1 in terms of the particular solution y_p and the functions $f(x) = e^x$ and $g(x) = e^{-2x}$.

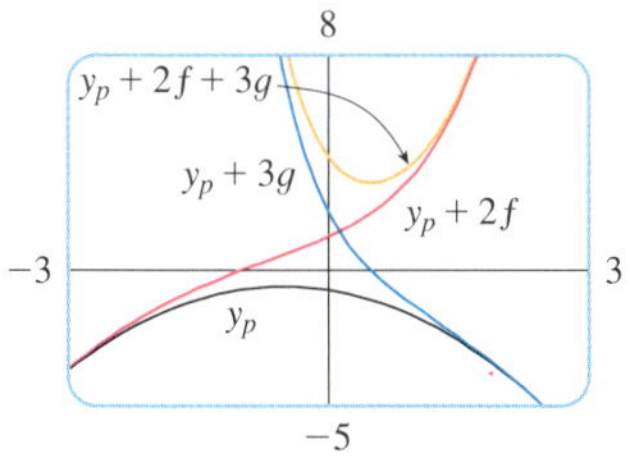

FIGURE 1

Polynomials are equal when their coefficients are equal. Thus

$$-2A = 1 \qquad 2A - 2B = 0 \qquad 2A + B - 2C = 0$$

The solution of this system of equations is

$$A = -\tfrac{1}{2} \qquad B = -\tfrac{1}{2} \qquad C = -\tfrac{3}{4}$$

A particular solution is therefore

$$y_p(x) = -\tfrac{1}{2}x^2 - \tfrac{1}{2}x - \tfrac{3}{4}$$

and, by Theorem 3, the general solution is

$$y = y_c + y_p = c_1e^x + c_2e^{-2x} - \tfrac{1}{2}x^2 - \tfrac{1}{2}x - \tfrac{3}{4}$$ ■

If $G(x)$ (the right side of Equation 1) is of the form Ce^{kx}, where C and k are constants, then we take as a trial solution a function of the same form, $y_p(x) = Ae^{kx}$, because the derivatives of e^{kx} are constant multiples of e^{kx}.

■ Figure 2 shows solutions of the differential equation in Example 2 in terms of y_p and the functions $f(x) = \cos 2x$ and $g(x) = \sin 2x$. Notice that all solutions approach ∞ as $x \to \infty$ and all solutions resemble sine functions when x is negative.

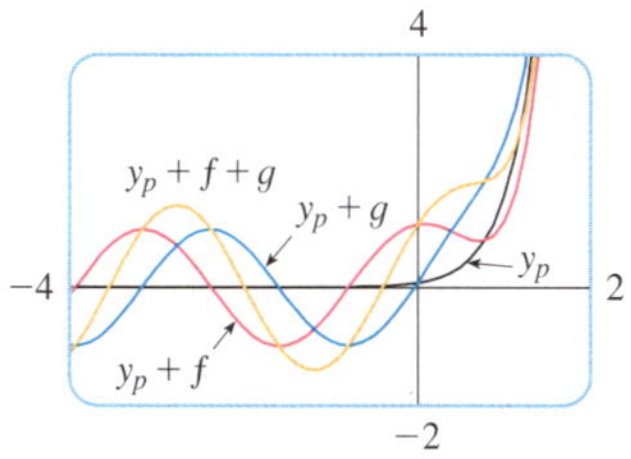

FIGURE 2

EXAMPLE 2 ■ Solve $y'' + 4y = e^{3x}$.

SOLUTION The auxiliary equation is $r^2 + 4 = 0$ with roots $\pm 2i$, so the solution of the complementary equation is

$$y_c(x) = c_1 \cos 2x + c_2 \sin 2x$$

For a particular solution we try $y_p(x) = Ae^{3x}$. Then $y_p' = 3Ae^{3x}$ and $y_p'' = 9Ae^{3x}$. Substituting into the differential equation, we have

$$9Ae^{3x} + 4(Ae^{3x}) = e^{3x}$$

so $13Ae^{3x} = e^{3x}$ and $A = \frac{1}{13}$. Thus, a particular solution is

$$y_p(x) = \tfrac{1}{13}e^{3x}$$

and the general solution is

$$y(x) = c_1 \cos 2x + c_2 \sin 2x + \tfrac{1}{13}e^{3x}$$ ■

If $G(x)$ is either $C \cos kx$ or $C \sin kx$, then, because of the rules for differentiating the sine and cosine functions, we take as a trial particular solution a function of the form

$$y_p(x) = A \cos kx + B \sin kx$$

EXAMPLE 3 ■ Solve $y'' + y' - 2y = \sin x$.

SOLUTION We try a particular solution

$$y_p(x) = A \cos x + B \sin x$$

Then $\qquad y_p' = -A \sin x + B \cos x \qquad y_p'' = -A \cos x - B \sin x$

so substitution in the differential equation gives

$$(-A \cos x - B \sin x) + (-A \sin x + B \cos x) - 2(A \cos x + B \sin x) = \sin x$$

or $$(-3A + B) \cos x + (-A - 3B) \sin x = \sin x$$

This is true if

$$-3A + B = 0 \qquad \text{and} \qquad -A - 3B = 1$$

The solution of this system is

$$A = -\tfrac{1}{10} \qquad B = -\tfrac{3}{10}$$

so a particular solution is

$$y_p(x) = -\tfrac{1}{10}\cos x - \tfrac{3}{10}\sin x$$

In Example 1 we determined that the solution of the complementary equation is $y_c = c_1e^x + c_2e^{-2x}$. Thus, the general solution of the given equation is

$$y(x) = c_1e^x + c_2e^{-2x} - \tfrac{1}{10}(\cos x + 3\sin x)$$ ■

If $G(x)$ is a product of functions of the preceding types, then we take the trial solution to be a product of functions of the same type. For instance, in solving the differential equation

$$y'' + 2y' + 4y = x\cos 3x$$

we would try

$$y_p(x) = (Ax + B)\cos 3x + (Cx + D)\sin 3x$$

If $G(x)$ is a sum of functions of these types, we use the easily verified *principle of superposition,* which says that if y_{p_1} and y_{p_2} are solutions of

$$ay'' + by' + cy = G_1(x) \qquad ay'' + by' + cy = G_2(x)$$

respectively, then $y_{p_1} + y_{p_2}$ is a solution of

$$ay'' + by' + cy = G_1(x) + G_2(x)$$

EXAMPLE 4 ■ Solve $y'' - 4y = xe^x + \cos 2x$.

SOLUTION The auxiliary equation is $r^2 - 4 = 0$ with roots ± 2, so the solution of the complementary equation is $y_c(x) = c_1e^{2x} + c_2e^{-2x}$. For the equation $y'' - 4y = xe^x$ we try

$$y_{p_1}(x) = (Ax + B)e^x$$

Then $y'_{p_1} = (Ax + A + B)e^x$, $y''_{p_1} = (Ax + 2A + B)e^x$, so substitution in the equation gives

$$(Ax + 2A + B)e^x - 4(Ax + B)e^x = xe^x$$

or

$$(-3Ax + 2A - 3B)e^x = xe^x$$

Thus, $-3A = 1$ and $2A - 3B = 0$, so $A = -\frac{1}{3}$, $B = -\frac{2}{9}$, and

$$y_{p_1}(x) = \left(-\tfrac{1}{3}x - \tfrac{2}{9}\right)e^x$$

■ In Figure 3 we show the particular solution y_p of the differential equation in Example 4. The other solutions are given in terms of $f(x) = e^{2x}$ and $g(x) = e^{-2x}$.

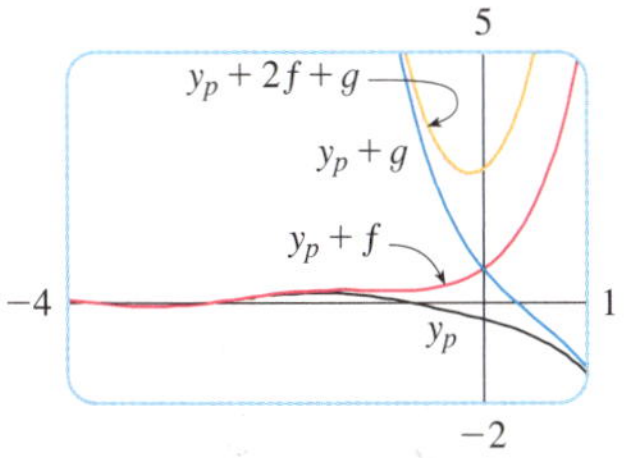

FIGURE 3

For the equation $y'' - 4y = \cos 2x$, we try

$$y_{p_2}(x) = C \cos 2x + D \sin 2x$$

Substitution gives

$$-4C \cos 2x - 4D \sin 2x - 4(C \cos 2x + D \sin 2x) = \cos 2x$$

or

$$-8C \cos 2x - 8D \sin 2x = \cos 2x$$

Therefore, $-8C = 1$, $-8D = 0$, and

$$y_{p_2}(x) = -\tfrac{1}{8} \cos 2x$$

By the superposition principle, the general solution is

$$y = y_c + y_{p_1} + y_{p_2} = c_1 e^{2x} + c_2 e^{-2x} - \left(\tfrac{1}{3}x + \tfrac{2}{9}\right)e^x - \tfrac{1}{8} \cos 2x$$ ■

Finally we note that the recommended trial solution y_p sometimes turns out to be a solution of the complementary equation and therefore can't be a solution of the nonhomogeneous equation. In such cases we multiply the recommended trial solution by x (or by x^2 if necessary) so that no term in $y_p(x)$ is a solution of the complementary equation.

EXAMPLE 5 ■ Solve $y'' + y = \sin x$.

SOLUTION The auxiliary equation is $r^2 + 1 = 0$ with roots $\pm i$, so the solution of the complementary equation is

$$y_c(x) = c_1 \cos x + c_2 \sin x$$

Ordinarily, we would use the trial solution

$$y_p(x) = A \cos x + B \sin x$$

but we observe that it is a solution of the complementary equation, so instead we try

$$y_p(x) = Ax \cos x + Bx \sin x$$

Then

$$y_p'(x) = A \cos x - Ax \sin x + B \sin x + Bx \cos x$$

$$y_p''(x) = -2A \sin x - Ax \cos x + 2B \cos x - Bx \sin x$$

■ The graphs of four solutions of the differential equation in Example 5 are shown in Figure 4.

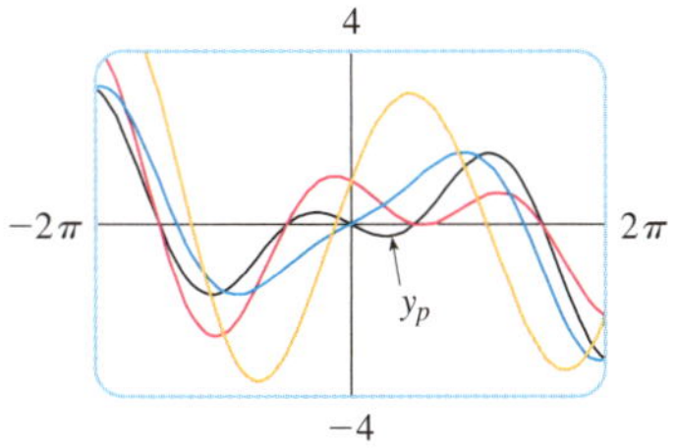

FIGURE 4

Substitution in the differential equation gives

$$y_p'' + y_p = -2A \sin x + 2B \cos x = \sin x$$

so $A = -\frac{1}{2}$, $B = 0$, and

$$y_p(x) = -\tfrac{1}{2}x \cos x$$

The general solution is

$$y(x) = c_1 \cos x + c_2 \sin x - \tfrac{1}{2}x \cos x$$ ■

Summary of Undetermined Coefficients

$G(x) =$	First try $y_p =$
$C_1x^n + \cdots + C_1x + C_0$	$A_1x^n + \cdots + A_1x + A_0$
Ce^{kx}	Ae^{kx}
$C\cos kx + D\sin kx$	$A\cos kx + B\sin kx$
Modification: If any term of y_p is a solution of the complementary equation, multiply y_p by x (or by x^2 if necessary).	

The Method of Variation of Parameters

Suppose we have already solved the homogeneous equation $ay'' + by' + cy = 0$ and written the solution as

4 $$y(x) = c_1y_1(x) + c_2y_2(x)$$

where y_1 and y_2 are linearly independent solutions. Let's replace the constants (or parameters) c_1 and c_2 in Equation 4 by arbitrary functions $u_1(x)$ and $u_2(x)$. We look for a particular solution of the nonhomogeneous equation $ay'' + by' + cy = G(x)$ of the form

5 $$y_p(x) = u_1(x)y_1(x) + u_2(x)y_2(x)$$

(This method is called **variation of parameters** because we have varied the parameters c_1 and c_2 to make them functions.) Differentiating Equation 5, we get

6 $$y_p' = (u_1'y_1 + u_2'y_2) + (u_1y_1' + u_2y_2')$$

Since u_1 and u_2 are arbitrary functions, we can impose two conditions on them. One condition is that y_p is a solution of the differential equation; we can choose the other condition so as to simplify our calculations. In view of the expression in Equation 6, let's impose the condition that

7 $$u_1'y_1 + u_2'y_2 = 0$$

Then $$y_p'' = u_1'y_1' + u_2'y_2' + u_1y_1'' + u_2y_2''$$

Substituting in the differential equation, we get

$$a(u_1'y_1' + u_2'y_2' + u_1y_1'' + u_2y_2'') + b(u_1y_1' + u_2y_2') + c(u_1y_1 + u_2y_2) = G$$

or

8 $$u_1(ay_1'' + by_1' + cy_1) + u_2(ay_2'' + by_2' + cy_2) + a(u_1'y_1' + u_2'y_2') = G$$

But y_1 and y_2 are solutions of the complementary equation, so

$$ay_1'' + by_1' + cy_1 = 0 \qquad \text{and} \qquad ay_2'' + by_2' + cy_2 = 0$$

and Equation 8 simplifies to

9 $$a(u_1'y_1' + u_2'y_2') = G$$

Equations 7 and 9 form a system of two equations in the unknown functions u_1' and u_2'. After solving this system we may be able to integrate to find u_1 and u_2 and then the particular solution is given by Equation 5.

EXAMPLE 6 ■ Solve the equation $y'' + y = \tan x$, $0 < x < \pi/2$.

SOLUTION The auxiliary equation is $r^2 + 1 = 0$ with roots $\pm i$, so the solution of $y'' + y = 0$ is $c_1 \sin x + c_2 \cos x$. Using variation of parameters, we seek a solution of the form

$$y_p(x) = u_1(x) \sin x + u_2(x) \cos x$$

Then $$y_p' = (u_1' \sin x + u_2' \cos x) + (u_1 \cos x - u_2 \sin x)$$

Set

10 $$u_1' \sin x + u_2' \cos x = 0$$

Then $$y_p'' = u_1' \cos x - u_2' \sin x - u_1 \sin x - u_2 \cos x$$

For y_p to be a solution we must have

11 $$y_p'' + y_p = u_1' \cos x - u_2' \sin x = \tan x$$

Solving Equations 10 and 11, we get

$$u_1'(\sin^2 x + \cos^2 x) = \cos x \tan x$$

$$u_1' = \sin x \qquad u_1(x) = -\cos x$$

(We seek a particular solution, so we don't need a constant of integration here.) Then, from Equation 10, we obtain

$$u_2' = -\frac{\sin x}{\cos x} u_1' = -\frac{\sin^2 x}{\cos x} = \frac{\cos^2 x - 1}{\cos x} = \cos x - \sec x$$

■ Figure 5 shows four solutions of the differential equation in Example 6.

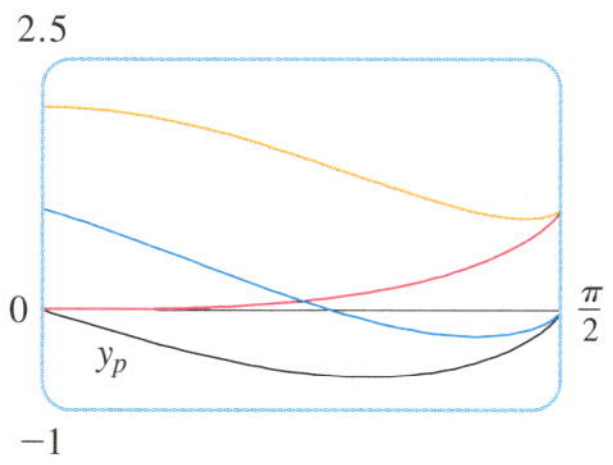

FIGURE 5

So $$u_2(x) = \sin x - \ln(\sec x + \tan x)$$

(Note that $\sec x + \tan x > 0$ for $0 < x < \pi/2$.) Therefore

$$\begin{aligned} y_p(x) &= -\cos x \sin x + (\sin x - \ln(\sec x + \tan x)) \cos x \\ &= -\cos x \ln(\sec x + \tan x) \end{aligned}$$

and the general solution is

$$y(x) = c_1 \sin x + c_2 \cos x - \cos x \ln(\sec x + \tan x)$$ ■

7.8 Exercises

1–10 ■ Solve the differential equation or initial-value problem using the method of undetermined coefficients.

1. $y'' + 3y' + 2y = x^2$

2. $y'' + 9y = e^{3x}$

3. $y'' - 2y' = \sin 4x$

4. $y'' + 6y' + 9y = 1 + x$

5. $y'' - 4y' + 5y = e^{-x}$

6. $y'' + 2y' + y = xe^{-x}$

7. $y'' + y = e^x + x^3$, $y(0) = 2$, $y'(0) = 0$

8. $y'' - 4y = e^x \cos x$, $y(0) = 1$, $y'(0) = 2$

9. $y'' - y = xe^{3x}$, $y(0) = 0$, $y'(0) = 1$

10. $y'' + y' - 2y = x + \sin 2x$, $y(0) = 1$, $y'(0) = 0$

11–12 ■ Graph the particular solution and several other solutions. What characteristics do these solutions have in common?

11. $4y'' + 5y' + y = e^x$

12. $2y'' + 3y' + y = 1 + \cos 2x$

13–16 ■ Write a trial solution for the method of undetermined coefficients. Do not determine the coefficients.

13. $y'' + 2y' + 6y = x^4 e^{2x}$

14. $y'' + 6y' + 2y = x^3 + e^x \sin 2x$

15. $y'' - 2y' + 2y = e^x \cos x$

16. $y'' + 3y' = 1 + xe^{-3x}$

17–20 ■ Solve the differential equation using (a) undetermined coefficients and (b) variation of parameters.

17. $y'' + 4y = x$

18. $y'' - 3y' + 2y = \sin x$

19. $y'' - 2y' + y = e^{2x}$

20. $y'' - y' = e^x$

21–26 ■ Solve the differential equation using the method of variation of parameters.

21. $y'' + y = \sec x$, $0 < x < \pi/2$

22. $y'' + y = \cot x$, $0 < x < \pi/2$

23. $y'' - 3y' + 2y = \dfrac{1}{1 + e^{-x}}$

24. $y'' + 3y' + 2y = \sin(e^x)$

25. $y'' - y = \dfrac{1}{x}$

26. $y'' + 4y' + 4y = \dfrac{e^{-2x}}{x^3}$

7.9 Applications of Second-Order Differential Equations

Second-order linear differential equations have a variety of applications in science and engineering. In this section we explore two of them: the vibration of springs and electric circuits.

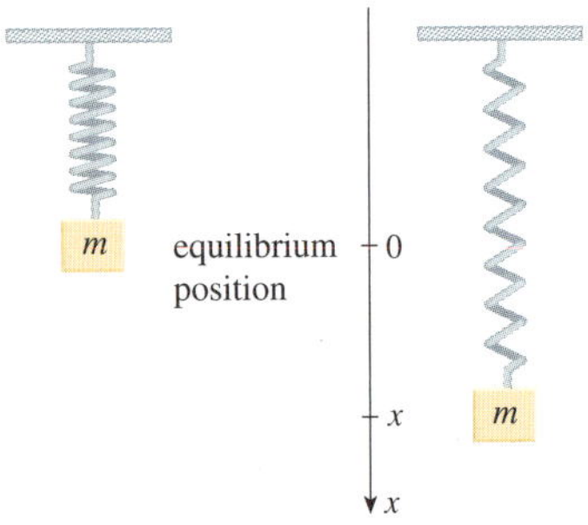

FIGURE 1

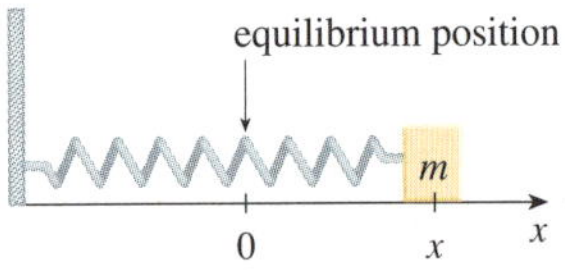

FIGURE 2

Vibrating Springs

We consider the motion of an object with mass m at the end of a spring that is either vertical (as in Figure 1) or horizontal on a level surface (as in Figure 2).

In Section 6.4 we discussed Hooke's Law, which says that if the spring is stretched (or compressed) x units from its natural length, then it exerts a force that is proportional to x:

$$\text{restoring force} = -kx$$

where k is a positive constant (called the **spring constant**). If we ignore any external resisting forces (due to air resistance or friction) then, by Newton's Second Law (force equals mass times acceleration), we have

$$\boxed{1} \qquad m\frac{d^2x}{dt^2} = -kx \qquad \text{or} \qquad m\frac{d^2x}{dt^2} + kx = 0$$

This is a second-order linear differential equation. Its auxiliary equation is $mr^2 + k = 0$

with roots $r = \pm\omega i$, where $\omega = \sqrt{k/m}$. Thus, the general solution is

$$x(t) = c_1 \cos \omega t + c_2 \sin \omega t$$

which can also be written as

$$x(t) = A \cos(\omega t + \delta)$$

where
$$\omega = \sqrt{k/m} \quad \text{(frequency)}$$
$$A = \sqrt{c_1^2 + c_2^2} \quad \text{(amplitude)}$$
$$\cos \delta = \frac{c_1}{A} \qquad \sin \delta = -\frac{c_2}{A} \qquad (\delta \text{ is the phase angle})$$

(See Exercise 15.) This type of motion is called **simple harmonic motion**.

EXAMPLE 1 ■ A spring with a mass of 2 kg has natural length 0.5 m. A force of 25.6 N is required to maintain it stretched to a length of 0.7 m. If the spring is stretched to a length of 0.7 m and then released with initial velocity 0, find the position of the mass at any time t.

SOLUTION From Hooke's Law, the force required to stretch the spring is

$$k(0.2) = 25.6$$

so $k = 25.6/0.2 = 128$. Using this value of the spring constant k, together with $m = 2$ in Equation 1, we have

$$2\frac{d^2x}{dt^2} + 128x = 0$$

As in the earlier general discussion, the solution of this equation is

$$\boxed{2} \qquad x(t) = c_1 \cos 8t + c_2 \sin 8t$$

We are given the initial condition that $x(0) = 0.2$. But, from Equation 2, $x(0) = c_1$. Therefore, $c_1 = 0.2$. Differentiating Equation 2, we get

$$x'(t) = -8c_1 \sin 8t + 8c_2 \cos 8t$$

Since the initial velocity is given as $x'(0) = 0$, we have $c_2 = 0$ and so the solution is

$$x(t) = \tfrac{1}{5} \cos 8t$$ ■

FIGURE 3

Damped Vibrations

We next consider the motion of a spring that is subject to a frictional force (in the case of the horizontal spring of Figure 2) or a damping force (in the case where a vertical spring moves through a fluid as in Figure 3). An example is the damping force supplied by a shock absorber in a car or a bicycle (see the photograph on page 1158).

We assume that the damping force is proportional to the velocity of the mass and acts in the direction opposite to the motion. (This has been confirmed, at least approximately, by some physical experiments.) Thus

$$\text{damping force} = -c\frac{dx}{dt}$$

where c is a positive constant, called the **damping constant**. Thus, in this case, Newton's Second Law gives

$$m\frac{d^2x}{dt^2} = \text{restoring force} + \text{damping force} = -kx - c\frac{dx}{dt}$$

or

3
$$m\frac{d^2x}{dt^2} + c\frac{dx}{dt} + kx = 0$$

Equation 3 is a second-order linear differential equation and its auxiliary equation is $mr^2 + cr + k = 0$. The roots are

4
$$r_1 = \frac{-c + \sqrt{c^2 - 4mk}}{2m} \qquad r_2 = \frac{-c - \sqrt{c^2 - 4mk}}{2m}$$

According to Section 18.1 we need to discuss three cases.

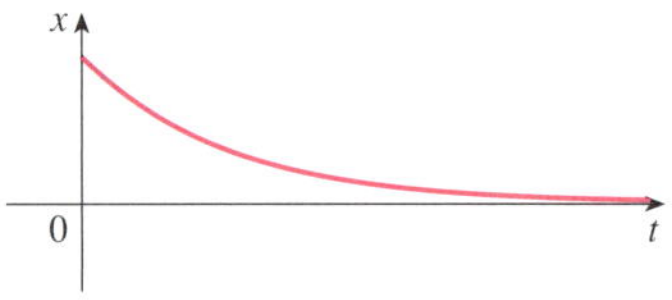

(a) c_1 and c_2 are positive.

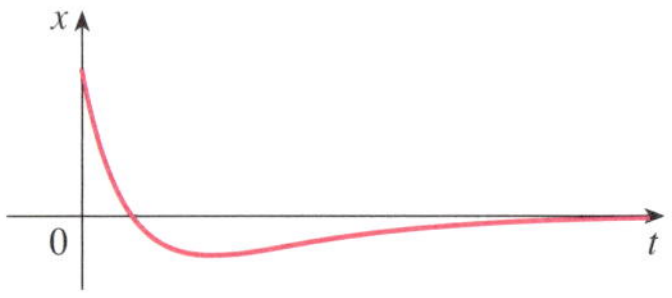

(b) c_1 and c_2 have opposite signs.

FIGURE 4
Overdamping

CASE I ■ $c^2 - 4mk > 0$ (overdamping)
In this case r_1 and r_2 are distinct real roots and

$$x = c_1e^{r_1t} + c_2e^{r_2t}$$

Since c, m, and k are all positive, we have $\sqrt{c^2 - 4mk} < c$, so the roots r_1 and r_2 given by Equations 4 must both be negative. This shows that $x \to 0$ as $t \to \infty$. Typical graphs of x as a function of t are shown in Figure 4. Notice that oscillations do not occur. This is because $c^2 > 4mk$ means that there is a strong damping force (high-viscosity oil or grease) compared with a weak spring or small mass.

CASE II ■ $c^2 - 4mk = 0$ (critical damping)
This case corresponds to equal roots

$$r_1 = r_2 = -\frac{c}{2m}$$

The solution is given by

$$x = (c_1 + c_2t)e^{-(c/2m)t}$$

and a typical graph is shown in Figure 5. It is similar to Case I, but the damping is just sufficient to suppress vibrations. Any decrease in the viscosity of the fluid leads to the vibrations of the following case.

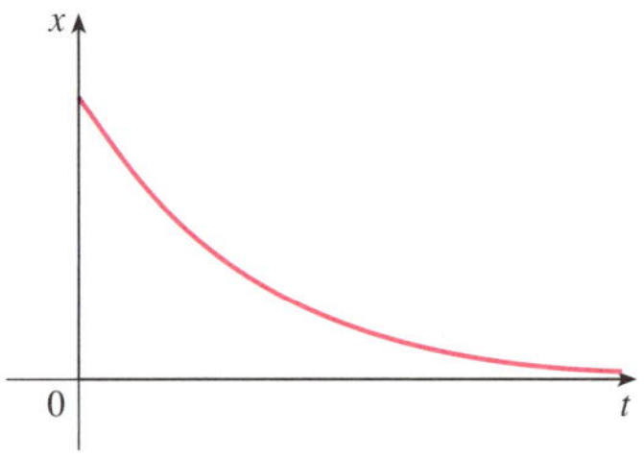

FIGURE 5
Critical damping

CASE III ■ $c^2 - 4mk < 0$ (underdamping)
Here the roots are complex:

$$\left.\begin{matrix} r_1 \\ r_2 \end{matrix}\right\} = -\frac{c}{2m} \pm \omega i$$

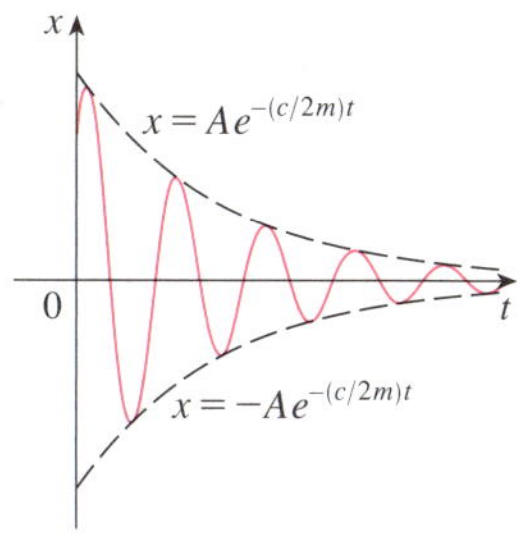

FIGURE 6
Underdamping

where
$$\omega = \frac{\sqrt{4mk - c^2}}{2m}$$

The solution is given by

$$x = e^{-(c/2m)t}(c_1 \cos \omega t + c_2 \sin \omega t)$$

We see that there are oscillations that are damped by the factor $e^{-(c/2m)t}$. Since $c > 0$ and $m > 0$, we have $-(c/2m) < 0$ so $e^{-(c/2m)t} \to 0$ as $t \to \infty$. This implies that $x \to 0$ as $t \to \infty$; that is, the motion decays to 0 as time increases. A typical graph is shown in Figure 6.

EXAMPLE 2 ■ Suppose that the spring of Example 1 is immersed in a fluid with damping constant $c = 40$. Find the position of the mass at any time t if it starts from the equilibrium position and is given a push to start it with an initial velocity of 0.6 m/s.

SOLUTION From Example 1 the mass is $m = 2$ and the spring constant is $k = 128$, so the differential equation (3) becomes

$$2\frac{d^2x}{dt^2} + 40\frac{dx}{dt} + 128x = 0$$

or
$$\frac{d^2x}{dt^2} + 20\frac{dx}{dt} + 64x = 0$$

The auxiliary equation is $r^2 + 20r + 64 = (r + 4)(r + 16) = 0$ with roots -4 and -16, so the motion is overdamped and the solution is

■ Figure 7 shows the graph of the position function for the overdamped motion in Example 2.

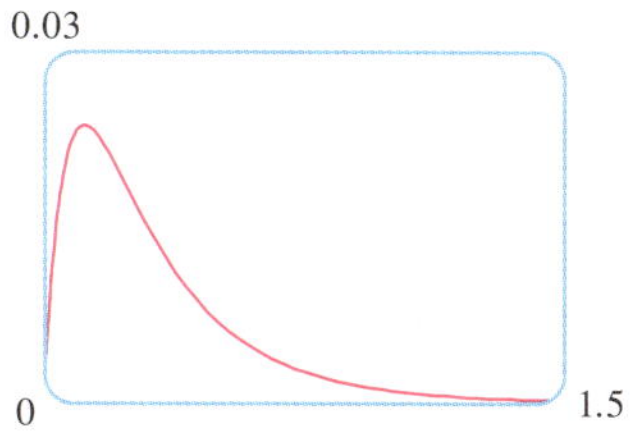

FIGURE 7

$$x(t) = c_1e^{-4t} + c_2e^{-16t}$$

We are given that $x(0) = 0$, so $c_1 + c_2 = 0$. Differentiating, we get

$$x'(t) = -4c_1e^{-4t} - 16c_2e^{-16t}$$

so
$$x'(0) = -4c_1 - 16c_2 = 0.6$$

Since $c_2 = -c_1$, this gives $12c_1 = 0.6$ or $c_1 = 0.05$. Therefore

$$x = 0.05(e^{-4t} - e^{-16t})$$ ■

Forced Vibrations

Suppose that, in addition to the restoring force and the damping force, the motion of the spring is affected by an external force $F(t)$. Then Newton's Second Law gives

$$m\frac{d^2x}{dt^2} = \text{restoring force} + \text{damping force} + \text{external force}$$

$$= -kx - c\frac{dx}{dt} + F(t)$$

Thus, instead of the homogeneous equation (3), the motion of the spring is now governed by the following nonhomogeneous differential equation:

$$m\frac{d^2x}{dt^2} + c\frac{dx}{dt} + kx = F(t) \tag{5}$$

The motion of the spring can be determined by the methods of Section 18.2.

A commonly occurring type of external force is a periodic force function

$$F(t) = F_0 \cos \omega_0 t \qquad \text{where} \quad \omega_0 \neq \omega = \sqrt{k/m}$$

In this case, and in the absence of a damping force ($c = 0$), you are asked in Exercise 9 to use the method of undetermined coefficients to show that

$$x(t) = c_1 \cos \omega t + c_2 \sin \omega t + \frac{F_0}{m(\omega^2 - \omega_0^2)} \cos \omega_0 t \tag{6}$$

If $\omega_0 = \omega$, then the applied frequency reinforces the natural frequency and the result is vibrations of large amplitude. This is the phenomenon of **resonance** (see Exercise 10).

Electric Curcuits

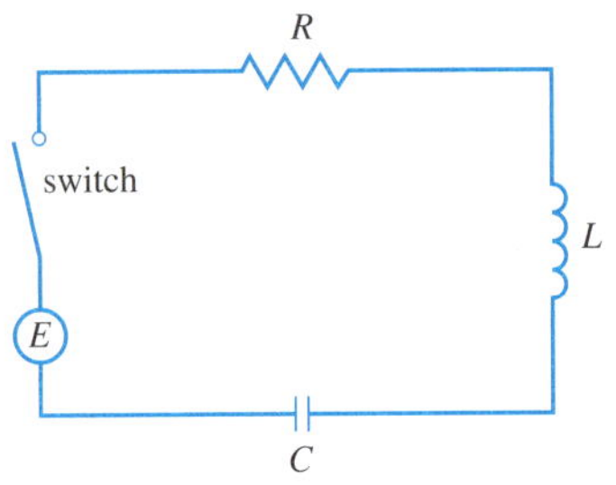

FIGURE 8

In Sections 10.3 and 10.6 we were able to use first-order separable and linear equations to analyze electric circuits that contain a resistor and inductor (see Figure 4 on page 631 or page 659) or a resistor and capacitor (see Exercise 29 on page 661). Now that we know how to solve second-order linear equations, we are in a position to analyze the circuit shown in Figure 8. It contains an electromotive force E (supplied by a battery or generator), a resistor R, an inductor L, and a capacitor C, in series. If the charge on the capacitor at time t is $Q = Q(t)$, then the current is the rate of change of Q with respect to t: $I = dQ/dt$. As in Section 10.6, it is known from physics that the voltage drops across the resistor, inductor, and capacitor are

$$RI \qquad L\frac{dI}{dt} \qquad \frac{Q}{C}$$

respectively. Kirchhoff's voltage law says that the sum of these voltage drops is equal to the supplied voltage:

$$L\frac{dI}{dt} + RI + \frac{Q}{C} = E(t)$$

Since $I = dQ/dt$, this equation becomes

$$L\frac{d^2Q}{dt^2} + R\frac{dQ}{dt} + \frac{1}{C}Q = E(t) \tag{7}$$

which is a second-order linear differential equation with constant coefficients. If the charge Q_0 and the current I_0 are known at time 0, then we have the initial conditions

$$Q(0) = Q_0 \qquad Q'(0) = I(0) = I_0$$

and the initial-value problem can be solved by the methods of Section 18.2.

7.9 Exercises

1. A spring with a 3-kg mass is held stretched 0.6 m beyond its natural length by a force of 20 N. If the spring begins at its equilibrium position but a push gives it an initial velocity of 1.2 m/s, find the position of the mass after t seconds.

2. A spring with a 4-kg mass has natural length 1 m and is maintained stretched to a length of 1.3 m by a force of 24.3 N. If the spring is compressed to a length of 0.8 m and then released with zero velocity, find the position of the mass at any time t.

3. A spring with a mass of 2 kg has damping constant 14, and a force of 6 N is required to keep the spring stretched 0.5 m beyond its natural length. The spring is stretched 1 m beyond its natural length and then released with zero velocity. Find the position of the mass at any time t.

4. A spring with a mass of 3 kg has damping constant 30 and spring constant 123.
   (a) Find the position of the mass at time t if it starts at the equilibrium position with a velocity of 2 m/s.
   (b) Graph the position function of the mass.

5. For the spring in Exercise 3, find the mass that would produce critical damping.

6. For the spring in Exercise 4, find the damping constant that would produce critical damping.

7. A spring has a mass of 1 kg and its spring constant is $k = 100$. The spring is released at a point 0.1 m above its equilibrium position. Graph the position function for the following values of the damping constant c: 10, 15, 20, 25, 30. What type of damping occurs in each case?

8. A spring has a mass of 1 kg and its damping constant is $c = 10$. The spring starts from its equilibrium position with a velocity of 1 m/s. Graph the position function for the following values of the spring constant k: 10, 20, 25, 30, 40. What type of damping occurs in each case?

9. Suppose a spring has mass m and spring constant k and let $\omega = \sqrt{k/m}$. Suppose that the damping constant is so small that the damping force is negligible. If an external force $F(t) = F_0 \cos \omega_0 t$ is applied, where $\omega_0 \neq \omega$, use the method of undetermined coefficients to show that the motion of the mass is described by Equation 6.

10. As in Exercise 9, consider a spring with mass m, spring constant k, and damping constant $c = 0$, and let $\omega = \sqrt{k/m}$. If an external force $F(t) = F_0 \cos \omega t$ is applied (the applied frequency equals the natural frequency), use the method of undetermined coefficients to show that the motion of the mass is given by $x(t) = c_1 \cos \omega t + c_2 \sin \omega t + (F_0/(2m\omega))t \sin \omega t$.

11. A series circuit consists of a resistor with $R = 20\ \Omega$, an inductor with $L = 1$ H, a capacitor with $C = 0.002$ F, and a 12-V battery. If the initial charge and current are both 0, find the charge and current at time t.

12. A series circuit contains a resistor with $R = 24\ \Omega$, an inductor with $L = 2$ H, a capacitor with $C = 0.005$ F, and a 12-V battery. The initial charge is $Q = 0.001$ C and the initial current is 0.
    (a) Find the charge and current at time t.
    (b) Graph the charge and current functions.

13. The battery in Exercise 11 is replaced by a generator producing a voltage of $E(t) = 12 \sin 10t$. Find the charge at time t.

14. The battery in Exercise 12 is replaced by a generator producing a voltage of $E(t) = 12 \sin 10t$.
    (a) Find the charge at time t.
    (b) Graph the charge function.

15. Verify that the solution to Equation 1 can be written in the form $x(t) = A \cos(\omega t + \delta)$.

16. The figure shows a pendulum with length L and the angle θ from the vertical to the pendulum. It can be shown that θ, as a function of time, satisfies the nonlinear differential equation

$$\frac{d^2\theta}{dt^2} + \frac{g}{L} \sin \theta = 0$$

where g is the acceleration due to gravity. For small values of θ we can use the linear approximation $\sin \theta \approx \theta$ and then the differential equation becomes linear.
    (a) Find the equation of motion of a pendulum with length 1 m if θ is initially 0.2 rad and the initial angular velocity is $d\theta/dt = 1$ rad/s.
    (b) What is the maximum angle from the vertical?
    (c) What is the period of the pendulum (that is, the time to complete one back-and-forth swing)?
    (d) When will the pendulum first be vertical?
    (e) What is the angular velocity when the pendulum is vertical?

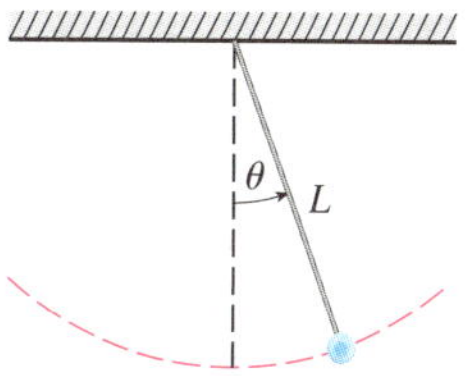

Chapter 18

Exercises 7.7 ■ page 1165

1. $y = c_1e^{4x} + c_2e^{2x}$ **3.** $y = e^{-4x}(c_1 \cos 5x + c_2 \sin 5x)$

5. $y = c_1e^{x} + c_2xe^{x}$ **7.** $y = c_1 \cos(x/2) + c_2 \sin(x/2)$

9. $y = c_1 + c_2e^{-x/4}$ **11.** $y = c_1e^{(1+\sqrt{2})t} + c_2e^{(1-\sqrt{2})t}$

13. $y = e^{-t/2}\left[c_1 \cos(\sqrt{3}t/2) + c_2 \sin(\sqrt{3}t/2)\right]$

15.

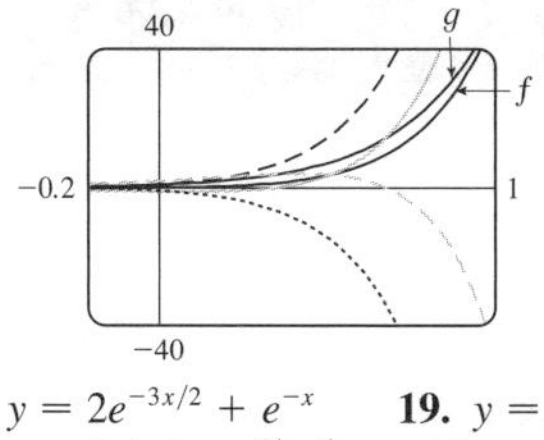

All solutions approach 0 as $x \to -\infty$ and approach $\pm\infty$ as $x \to \infty$.

17. $y = 2e^{-3x/2} + e^{-x}$ **19.** $y = e^{x}(\cos x + \sin x)$

21. $y = 2e^{1-x} + e^{3(x-1)}$ **23.** $y = -\frac{1}{3} \sin 3x$

25. $y = 3xe^{-2x+2}$ **27.** No solution

29. $y = \dfrac{e^5}{e^6 - 1}e^{-x} + \dfrac{e^2}{1 - e^6}e^{2x}$

31. $y = e^{-2x}(2 \cos 3x - e^{\pi} \sin 3x)$

33. (b) $\lambda = n^2\pi^2/L^2$, n a positive integer; $y = C \sin(n\pi x/L)$

Exercises 7.8 ■ page 1172

1. $y = c_1e^{-2x} + c_2e^{-x} + \frac{1}{2}x^2 - \frac{3}{2}x + \frac{7}{4}$

3. $y = c_1 + c_2e^{2x} + \frac{1}{40} \cos 4x - \frac{1}{20} \sin 4x$

5. $y = e^{2x}(c_1 \cos x + c_2 \sin x) + \frac{1}{10}e^{-x}$

7. $y = \frac{3}{2} \cos x + \frac{11}{2} \sin x + \frac{1}{2}e^{x} + x^3 - 6x$

9. $y = \frac{5}{8}e^{x} - \frac{17}{32}e^{-x} + e^{3x}\left[\frac{1}{8}x - \frac{3}{32}\right]$

11.

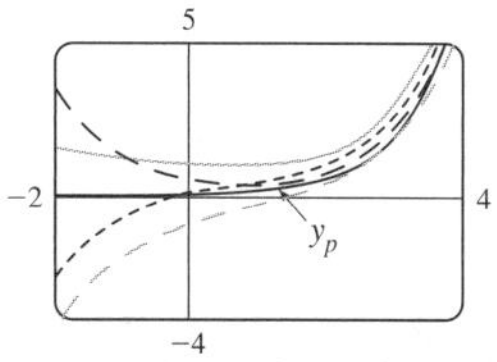

The solutions are all asymptotic to $y_p = e^{x}/10$ as $x \to \infty$. Except for y_p, all solutions approach either ∞ or $-\infty$ as $x \to -\infty$.

13. $y_p = (Ax^4 + Bx^3 + Cx^2 + Dx + E)e^{2x}$

15. $y_p = xe^{x}(A \cos x + B \sin x)$

17. $y = c_1 \cos 2x + c_2 \sin 2x + \frac{1}{4}x$

19. $y = c_1e^{x} + c_2xe^{x} + e^{2x}$

21. $y = (c_1 + x) \sin x + (c_2 + \ln \cos x) \cos x$

23. $y = [c_1 + \ln(1 + e^{-x})]e^{x} + [c_2 - e^{-x} + \ln(1 + e^{-x})]e^{2x}$

25. $y = \left[c_1 - \frac{1}{2}\int (e^{x}/x)\,dx\right]e^{-x} + \left[c_2 + \frac{1}{2}\int (e^{-x}/x)\,dx\right]e^{x}$

Exercises 7.9 ■ page 1179

1. $x = 0.36 \sin(10t/3)$ **3.** $x = -\frac{1}{5}e^{-6t} + \frac{6}{5}e^{-t}$ **5.** $\frac{49}{12}$ kg

7.

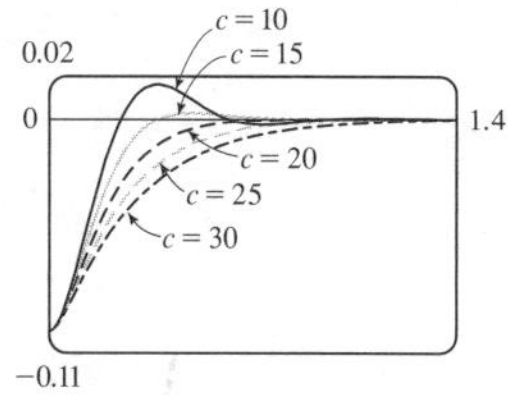

11. $Q(t) = (-e^{-10t}/250)(6 \cos 20t + 3 \sin 20t) + \frac{3}{125}$, $I(t) = \frac{3}{5}e^{-10t} \sin 20t$

13. $Q(t) = e^{-10t}\left[\frac{3}{250} \cos 20t - \frac{3}{500} \sin 20t\right] - \frac{3}{250} \cos 10t + \frac{3}{125} \sin 10t$

8 Infinite Sequences and Series

Everything that we have done in this section can be extended to functions of three or more variables. The notation

$$\lim_{(x, y, z) \to (a, b, c)} f(x, y, z) = L$$

means that the values of $f(x, y, z)$ approach the number L as the point (x, y, z) approaches the point (a, b, c) along any path in the domain of f. The function f is **continuous** at (a, b, c) if

$$\lim_{(x, y, z) \to (a, b, c)} f(x, y, z) = f(a, b, c)$$

For instance, the function

$$f(x, y, z) = \frac{1}{x^2 + y^2 + z^2 - 1}$$

is a rational function of three variables and so is continuous at every point in $\mathbb{R}^3$ except where $x^2 + y^2 + z^2 = 1$. In other words, it is discontinuous on the sphere with center the origin and radius 1.

11.2 Exercises

1. Suppose that $\lim_{(x, y) \to (3, 1)} f(x, y) = 6$. What can you say about the value of $f(3, 1)$? What if f is continuous?

2. Explain why each function is continuous or discontinuous.
   (a) The outdoor temperature as a function of longitude, latitude, and time
   (b) Elevation (height above sea level) as a function of longitude, latitude, and time
   (c) The cost of a taxi ride as a function of distance traveled and time

3–4 ■ Use a table of numerical values of $f(x, y)$ for (x, y) near the origin to make a conjecture about the value of the limit of $f(x, y)$ as $(x, y) \to (0, 0)$. Then explain why your guess is correct.

3. $f(x, y) = \dfrac{x^2y^3 + x^3y^2 - 5}{2 - xy}$

4. $f(x, y) = \dfrac{2xy}{x^2 + 2y^2}$

5–18 ■ Find the limit, if it exists, or show that the limit does not exist.

5. $\lim_{(x, y) \to (5, -2)} (x^5 + 4x^3y - 5xy^2)$

6. $\lim_{(x, y) \to (6, 3)} xy \cos(x - 2y)$

7. $\lim_{(x, y) \to (0, 0)} \dfrac{x^2}{x^2 + y^2}$

8. $\lim_{(x, y) \to (0, 0)} \dfrac{(x + y)^2}{x^2 + y^2}$

9. $\lim_{(x, y) \to (0, 0)} \dfrac{8x^2y^2}{x^4 + y^4}$

10. $\lim_{(x, y) \to (0, 0)} \dfrac{x^3 + xy^2}{x^2 + y^2}$

11. $\lim_{(x, y) \to (0, 0)} \dfrac{xy}{\sqrt{x^2 + y^2}}$

12. $\lim_{(x, y) \to (0, 0)} \dfrac{x^2 \sin^2 y}{x^2 + 2y^2}$

13. $\lim_{(x, y) \to (0, 0)} \dfrac{2x^2y}{x^4 + y^2}$

14. $\lim_{(x, y) \to (2, 0)} \dfrac{xy - 2y}{x^2 + y^2 - 4x + 4}$

15. $\lim_{(x, y) \to (0, 0)} \dfrac{x^2 + y^2}{\sqrt{x^2 + y^2 + 1} - 1}$

16. $\lim_{(x, y, z) \to (3, -2, 2)} e^{x^2 z} \cos(y + z)$

17. $\lim_{(x, y, z) \to (0, 0, 0)} \dfrac{xy + yz^2 + xz^2}{x^2 + y^2 + z^4}$

18. $\lim_{(x, y, z) \to (0, 0, 0)} \dfrac{x^2 + 2y^2 + 3z^2}{x^2 + y^2 + z^2}$

19–20 ■ Use a computer graph of the function to explain why the limit does not exist.

19. $\lim_{(x, y) \to (0, 0)} \dfrac{2x^2 + 3xy + 4y^2}{3x^2 + 5y^2}$

20. $\lim_{(x, y) \to (0, 0)} \dfrac{xy^3}{x^2 + y^6}$

21–22 ■ Find $h(x, y) = g(f(x, y))$ and the set on which h is continuous.

21. $g(t) = t^2 + \sqrt{t}$, $f(x, y) = 2x + 3y - 6$

22. $g(z) = \sin z$, $f(x, y) = y \ln x$

23–24 ■ Graph the function and observe where it is discontinuous. Then use the formula to explain what you have observed.

23. $f(x, y) = e^{1/(x-y)}$

24. $f(x, y) = \dfrac{1}{1 - x^2 - y^2}$

25–32 ■ Determine the largest set on which the function is continuous.

25. $F(x, y) = \dfrac{1}{x^2 - y}$

26. $F(x, y) = \dfrac{x - y}{1 + x^2 + y^2}$

27. $F(x, y) = \arctan(x + \sqrt{y})$

28. $G(x, y) = \sin^{-1}(x^2 + y^2)$

29. $f(x, y, z) = \dfrac{xyz}{x^2 + y^2 - z}$

30. $f(x, y, z) = \sqrt{x + y + z}$

31. $f(x, y) = \begin{cases} \dfrac{x^2y^3}{2x^2 + y^2} & \text{if } (x, y) \neq (0, 0) \\ 1 & \text{if } (x, y) = (0, 0) \end{cases}$

32. $f(x, y) = \begin{cases} \dfrac{xy}{x^2 + xy + y^2} & \text{if } (x, y) \neq (0, 0) \\ 0 & \text{if } (x, y) = (0, 0) \end{cases}$

33–34 ■ Use polar coordinates to find the limit. [If (r, θ) are polar coordinates of the point (x, y) with $r \geqslant 0$, note that $r \to 0^+$ as $(x, y) \to (0, 0)$.]

33. $\displaystyle\lim_{(x, y) \to (0, 0)} \frac{x^3 + y^3}{x^2 + y^2}$

34. $\displaystyle\lim_{(x, y) \to (0, 0)} (x^2 + y^2) \ln(x^2 + y^2)$

35. Use spherical coordinates to find

$$\lim_{(x, y, z) \to (0, 0, 0)} \frac{xyz}{x^2 + y^2 + z^2}$$

36. At the beginning of this section we considered the function

$$f(x, y) = \frac{\sin(x^2 + y^2)}{x^2 + y^2}$$

and guessed that $f(x, y) \to 1$ as $(x, y) \to (0, 0)$ on the basis of numerical evidence. Use polar coordinates to confirm the value of the limit. Then graph the function.

Limits and Continuity

Let's compare the behavior of the functions

$$f(x, y) = \frac{\sin(x^2 + y^2)}{x^2 + y^2} \qquad \text{and} \qquad g(x, y) = \frac{x^2 - y^2}{x^2 + y^2}$$

as x and y both approach 0 [and therefore the point (x, y) approaches the origin].

TABLE 1 Values of $f(x, y)$

$x \backslash y$	−1.0	−0.5	−0.2	0	0.2	0.5	1.0
−1.0	0.455	0.759	0.829	0.841	0.829	0.759	0.455
−0.5	0.759	0.959	0.986	0.990	0.986	0.959	0.759
−0.2	0.829	0.986	0.999	1.000	0.999	0.986	0.829
0	0.841	0.990	1.000		1.000	0.990	0.841
0.2	0.829	0.986	0.999	1.000	0.999	0.986	0.829
0.5	0.759	0.959	0.986	0.990	0.986	0.959	0.759
1.0	0.455	0.759	0.829	0.841	0.829	0.759	0.455

TABLE 2 Values of $g(x, y)$

$x \backslash y$	−1.0	−0.5	−0.2	0	0.2	0.5	1.0
−1.0	0.000	0.600	0.923	1.000	0.923	0.600	0.000
−0.5	−0.600	0.000	0.724	1.000	0.724	0.000	−0.600
−0.2	−0.923	−0.724	0.000	1.000	0.000	−0.724	−0.923
0	−1.000	−1.000	−1.000		−1.000	−1.000	−1.000
0.2	−0.923	−0.724	0.000	1.000	0.000	−0.724	−0.923
0.5	−0.600	0.000	0.724	1.000	0.724	0.000	−0.600
1.0	0.000	0.600	0.923	1.000	0.923	0.600	0.000

Tables 1 and 2 show values of $f(x, y)$ and $g(x, y)$, correct to three decimal places, for points (x, y) near the origin. (Notice that neither function is defined at the origin.) It appears that as (x, y) approaches $(0, 0)$, the values of $f(x, y)$ are approaching 1 whereas the values of $g(x, y)$ aren't approaching any number. It turns out that these guesses based on numerical evidence are correct, and we write

$$\lim_{(x, y)\to(0, 0)} \frac{\sin(x^2 + y^2)}{x^2 + y^2} = 1 \quad \text{and} \quad \lim_{(x, y)\to(0, 0)} \frac{x^2 - y^2}{x^2 + y^2} \quad \text{does not exist}$$

In general, we use the notation

$$\lim_{(x, y)\to(a, b)} f(x, y) = L$$

to indicate that the values of $f(x, y)$ approach the number L as the point (x, y) approaches the point (a, b) along any path that stays within the domain of f. In other words, we can make the values of $f(x, y)$ as close to L as we like by taking the point (x, y) sufficiently close to the point (a, b), but not equal to (a, b). A more precise definition follows.

1 Definition Let f be a function of two variables whose domain D includes points arbitrarily close to (a, b). Then we say that the **limit of $f(x, y)$ as (x, y) approaches (a, b)** is L and we write

$$\lim_{(x, y)\to(a, b)} f(x, y) = L$$

if for every number $\varepsilon > 0$ there is a corresponding number $\delta > 0$ such that

$$|f(x, y) - L| < \varepsilon \quad \text{whenever} \quad (x, y) \in D \quad \text{and} \quad 0 < \sqrt{(x - a)^2 + (y - b)^2} < \delta$$

Other notations for the limit in Definition 1 are

$$\lim_{\substack{x\to a \\ y\to b}} f(x, y) = L \qquad \text{and} \qquad f(x, y) \to L \text{ as } (x, y) \to (a, b)$$

Notice that $|f(x, y) - L|$ is the distance between the numbers $f(x, y)$ and L, and $\sqrt{(x - a)^2 + (y - b)^2}$ is the distance between the point (x, y) and the point (a, b). Thus, Definition 1 says that the distance between $f(x, y)$ and L can be made arbitrarily small by making the distance from (x, y) to (a, b) sufficiently small (but not 0). Figure 1 illustrates Definition 1 by means of an arrow diagram. If any small interval $(L - \varepsilon, L + \varepsilon)$ is given around L, then we can find a disk D_δ with center (a, b) and radius $\delta > 0$ such that f maps all the points in D_δ [except possibly (a, b)] into the interval $(L - \varepsilon, L + \varepsilon)$.

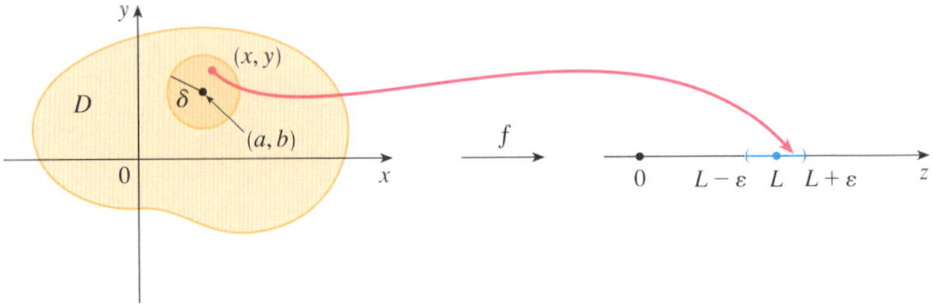

FIGURE 1

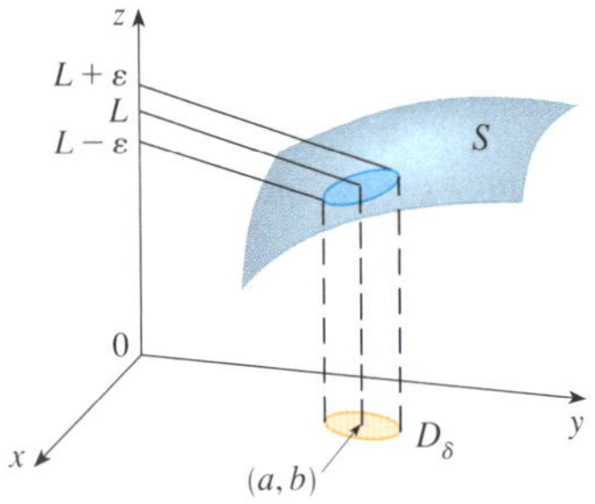

FIGURE 2

Another illustration of Definition 1 is given in Figure 2 where the surface S is the graph of f. If $\varepsilon > 0$ is given, we can find $\delta > 0$ such that if (x, y) is restricted to lie in the disk D_δ and $(x, y) \neq (a, b)$, then the corresponding part of S lies between the horizontal planes $z = L - \varepsilon$ and $z = L + \varepsilon$.

For functions of a single variable, when we let x approach a, there are only two possible directions of approach, from the left or from the right. We recall from Chapter 2 that if $\lim_{x\to a^-} f(x) \neq \lim_{x\to a^+} f(x)$, then $\lim_{x\to a} f(x)$ does not exist.

For functions of two variables the situation is not as simple because we can let (x, y) approach (a, b) from an infinite number of directions in any manner whatsoever (see Figure 3) as long as (x, y) stays within the domain of f.

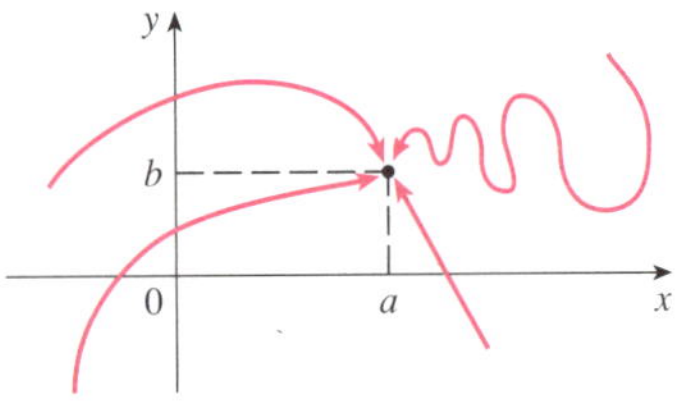

FIGURE 3

Definition 1 says that the distance between $f(x, y)$ and L can be made arbitrarily small by making the distance from (x, y) to (a, b) sufficiently small (but not 0). The definition refers only to the *distance* between (x, y) and (a, b). It does not refer to the direction of approach. Therefore, if the limit exists, then $f(x, y)$ must approach the same limit no matter how (x, y) approaches (a, b). Thus, if we can find two different paths of approach along which $f(x, y)$ has different limits, then it follows that $\lim_{(x, y)\to(a, b)} f(x, y)$ does not exist.

> If $f(x, y) \to L_1$ as $(x, y) \to (a, b)$ along a path C_1 and $f(x, y) \to L_2$ as $(x, y) \to (a, b)$ along a path C_2, where $L_1 \neq L_2$, then $\lim_{(x, y)\to(a, b)} f(x, y)$ does not exist.

EXAMPLE 1 □ Show that $\displaystyle\lim_{(x, y)\to(0, 0)} \frac{x^2 - y^2}{x^2 + y^2}$ does not exist.

SOLUTION Let $f(x, y) = (x^2 - y^2)/(x^2 + y^2)$. First let's approach $(0, 0)$ along the x-axis. Then $y = 0$ gives $f(x, 0) = x^2/x^2 = 1$ for all $x \neq 0$, so

$$f(x, y) \to 1 \qquad \text{as} \qquad (x, y) \to (0, 0) \text{ along the } x\text{-axis}$$

We now approach along the y-axis by putting $x = 0$. Then $f(0, y) = \dfrac{-y^2}{y^2} = -1$ for all $y \neq 0$, so

$$f(x, y) \to -1 \qquad \text{as} \qquad (x, y) \to (0, 0) \text{ along the } y\text{-axis}$$

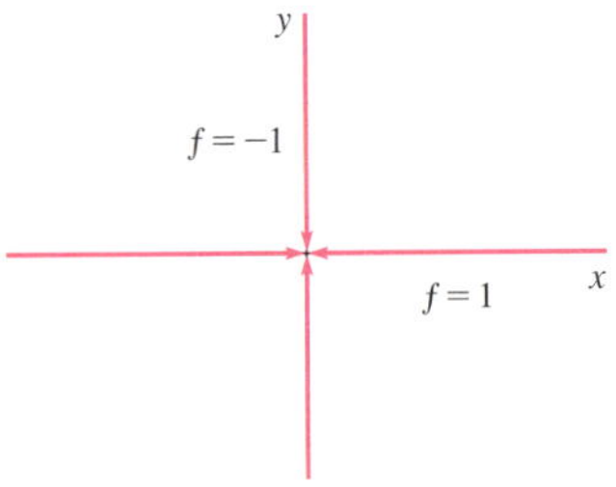

FIGURE 4

(See Figure 4.) Since f has two different limits along two different lines, the given limit does not exist. (This confirms the conjecture we made on the basis of numerical evidence at the beginning of this section.) □

EXAMPLE 2 □ If $f(x, y) = xy/(x^2 + y^2)$, does $\lim_{(x, y) \to (0, 0)} f(x, y)$ exist?

SOLUTION If $y = 0$, then $f(x, 0) = 0/x^2 = 0$. Therefore

$$f(x, y) \to 0 \quad \text{as} \quad (x, y) \to (0, 0) \text{ along the } x\text{-axis}$$

If $x = 0$, then $f(0, y) = 0/y^2 = 0$, so

$$f(x, y) \to 0 \quad \text{as} \quad (x, y) \to (0, 0) \text{ along the } y\text{-axis}$$

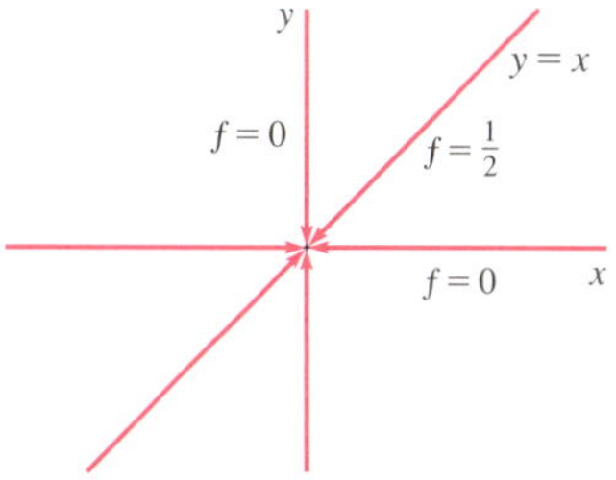

FIGURE 5

Although we have obtained identical limits along the axes, that does not show that the given limit is 0. Let's now approach (0, 0) along another line, say $y = x$. For all $x \neq 0$,

$$f(x, x) = \frac{x^2}{x^2 + x^2} = \frac{1}{2}$$

Therefore $\qquad f(x, y) \to \frac{1}{2} \quad \text{as} \quad (x, y) \to (0, 0) \text{ along } y = x$

(See Figure 5.) Since we have obtained different limits along different paths, the given limit does not exist. □

Figure 6 sheds some light on Example 2. The ridge that occurs above the line $y = x$ corresponds to the fact that $f(x, y) = \frac{1}{2}$ for all points (x, y) on that line except the origin.

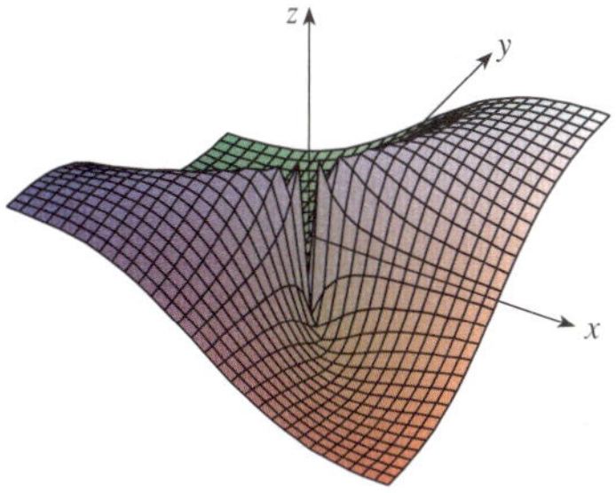

FIGURE 6
$f(x, y) = \dfrac{xy}{x^2 + y^2}$

EXAMPLE 3 □ If $f(x, y) = \dfrac{xy^2}{x^2 + y^4}$, does $\lim_{(x, y) \to (0, 0)} f(x, y)$ exist?

SOLUTION With the solution of Example 2 in mind, let's try to save time by letting $(x, y) \to (0, 0)$ along any nonvertical line through the origin. Then $y = mx$, where m is the slope, and

$$f(x, y) = f(x, mx) = \frac{x(mx)^2}{x^2 + (mx)^4} = \frac{m^2x^3}{x^2 + m^4x^4} = \frac{m^2x}{1 + m^4x^2}$$

So $\qquad f(x, y) \to 0 \quad \text{as} \quad (x, y) \to (0, 0) \text{ along } y = mx$

□ Figure 7 shows the graph of the function in Example 3. Notice the ridge above the parabola $x = y^2$.

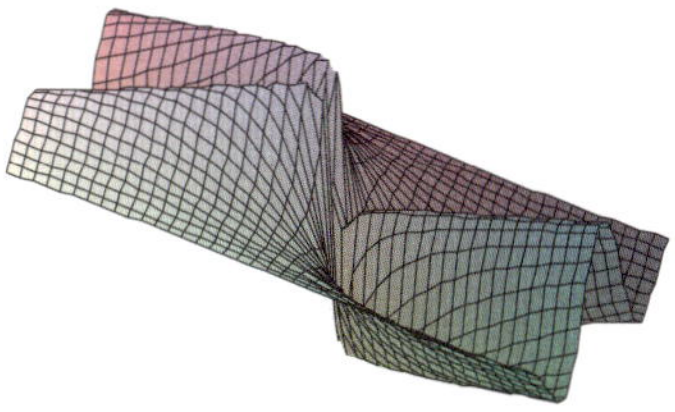

FIGURE 7

Thus, f has the same limiting value along every nonvertical line through the origin. But that does not show that the given limit is 0, for if we now let $(x, y) \to (0, 0)$ along the parabola $x = y^2$, we have

$$f(x, y) = f(y^2, y) = \frac{y^2 \cdot y^2}{(y^2)^2 + y^4} = \frac{y^4}{2y^4} = \frac{1}{2}$$

so $\qquad f(x, y) \to \frac{1}{2} \quad \text{as} \quad (x, y) \to (0, 0) \text{ along } x = y^2$

Since different paths lead to different limiting values, the given limit does not exist. □

Now let's look at limits that *do* exist. Just as for functions of one variable, the calculation of limits for functions of two variables can be greatly simplified by the use of properties of limits. The Limit Laws listed in Section 2.3 can be extended to functions of two variables. The limit of a sum is the sum of the limits, the limit of a product is the product of the limits, and so on. In particular, the following equations are true.

2
$$\lim_{(x,y)\to(a,b)} x = a \qquad \lim_{(x,y)\to(a,b)} y = b \qquad \lim_{(x,y)\to(a,b)} c = c$$

The Squeeze Theorem also holds.

EXAMPLE 4 □ Find $\displaystyle\lim_{(x,y)\to(0,0)} \frac{3x^2y}{x^2+y^2}$ if it exists.

SOLUTION As in Example 3, we could show that the limit along any line through the origin is 0. This doesn't prove that the given limit is 0, but the limits along the parabolas $y = x^2$ and $x = y^2$ also turn out to be 0, so we begin to suspect that the limit does exist and is equal to 0.

Let $\varepsilon > 0$. We want to find $\delta > 0$ such that

$$\left|\frac{3x^2y}{x^2+y^2} - 0\right| < \varepsilon \qquad \text{whenever} \qquad 0 < \sqrt{x^2+y^2} < \delta$$

that is,
$$\frac{3x^2|y|}{x^2+y^2} < \varepsilon \qquad \text{whenever} \qquad 0 < \sqrt{x^2+y^2} < \delta$$

But $x^2 \leqslant x^2 + y^2$ since $y^2 \geqslant 0$, so $x^2/(x^2+y^2) \leqslant 1$ and therefore

3
$$\frac{3x^2|y|}{x^2+y^2} \leqslant 3|y| = 3\sqrt{y^2} \leqslant 3\sqrt{x^2+y^2}$$

Thus, if we choose $\delta = \varepsilon/3$ and let $0 < \sqrt{x^2+y^2} < \delta$, then

$$\left|\frac{3x^2y}{x^2+y^2} - 0\right| \leqslant 3\sqrt{x^2+y^2} \leqslant 3\delta = 3\left(\frac{\varepsilon}{3}\right) = \varepsilon$$

Hence, by Definition 1,

$$\lim_{(x,y)\to(0,0)} \frac{3x^2y}{x^2+y^2} = 0$$ □

□ Another way to do Example 4 is to use the Squeeze Theorem instead of Definition 1. From (2) it follows that

$$\lim_{(x,y)\to(0,0)} 3|y| = 0$$

and so the first inequality in (3) shows that the given limit is 0.

Continuity

Recall that evaluating limits of *continuous* functions of a single variable is easy. It can be accomplished by direct substitution because the defining property of a continuous function is $\lim_{x\to a} f(x) = f(a)$. Continuous functions of two variables are also defined by the direct substitution property.

4 Definition A function f of two variables is called **continuous at** (a, b) if

$$\lim_{(x,y)\to(a,b)} f(x,y) = f(a,b)$$

We say f is **continuous on** D if f is continuous at every point (a, b) in D.

The intuitive meaning of continuity is that if the point (x, y) changes by a small amount, then the value of $f(x, y)$ changes by a small amount. This means that a surface that is the graph of a continuous function has no hole or break.

Using the properties of limits, you can see that sums, differences, products, and quotients of continuous functions are continuous on their domains. Let's use this fact to give examples of continuous functions.

A **polynomial function of two variables** (or polynomial, for short) is a sum of terms of the form cx^my^n, where c is a constant and m and n are nonnegative integers. A **rational function** is a ratio of polynomials. For instance,

$$f(x, y) = x^4 + 5x^3y^2 + 6xy^4 - 7y + 6$$

is a polynomial, whereas

$$g(x, y) = \frac{2xy + 1}{x^2 + y^2}$$

is a rational function.

The limits in (2) show that the functions $f(x, y) = x$, $g(x, y) = y$, and $h(x, y) = c$ are continuous. Since any polynomial can be built up out of the simple functions f, g, and h by multiplication and addition, it follows that *all polynomials are continuous on* $\mathbb{R}^2$. Likewise, any rational function is continuous on its domain because it is a quotient of continuous functions.

EXAMPLE 5 □ Evaluate $\displaystyle\lim_{(x, y)\to(1, 2)} (x^2y^3 - x^3y^2 + 3x + 2y)$.

SOLUTION Since $f(x, y) = x^2y^3 - x^3y^2 + 3x + 2y$ is a polynomial, it is continuous everywhere, so we can find the limit by direct substitution:

$$\lim_{(x, y)\to(1, 2)} (x^2y^3 - x^3y^2 + 3x + 2y) = 1^2 \cdot 2^3 - 1^3 \cdot 2^2 + 3 \cdot 1 + 2 \cdot 2 = 11$$ □

EXAMPLE 6 □ Where is the function $f(x, y) = \dfrac{x^2 - y^2}{x^2 + y^2}$ continuous?

SOLUTION The function f is discontinuous at $(0, 0)$ because it is not defined there. Since f is a rational function, it is continuous on its domain, which is the set $D = \{(x, y) \mid (x, y) \neq (0, 0)\}$. □

EXAMPLE 7 □ Let

$$g(x, y) = \begin{cases} \dfrac{x^2 - y^2}{x^2 + y^2} & \text{if } (x, y) \neq (0, 0) \\ 0 & \text{if } (x, y) = (0, 0) \end{cases}$$

Here g is defined at $(0, 0)$ but g is still discontinuous at 0 because $\lim_{(x, y)\to(0, 0)} g(x, y)$ does not exist (see Example 1). □

EXAMPLE 8 □ Let

$$f(x, y) = \begin{cases} \dfrac{3x^2y}{x^2 + y^2} & \text{if } (x, y) \neq (0, 0) \\ 0 & \text{if } (x, y) = (0, 0) \end{cases}$$

□ Figure 8 shows the graph of the continuous function in Example 8.

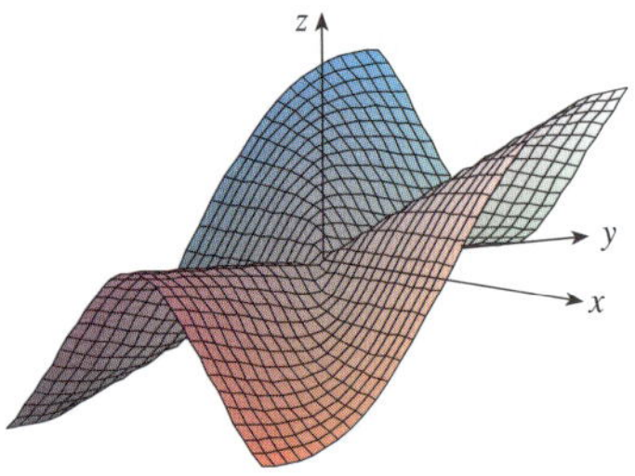

FIGURE 8

We know f is continuous for $(x, y) \neq (0, 0)$ since it is equal to a rational function there. Also, from Example 4, we have

$$\lim_{(x, y) \to (0, 0)} f(x, y) = \lim_{(x, y) \to (0, 0)} \frac{3x^2y}{x^2 + y^2} = 0 = f(0, 0)$$

Therefore, f is continuous at $(0, 0)$, and so it is continuous on $\mathbb{R}^2$. □

Just as for functions of one variable, composition is another way of combining two continuous functions to get a third. In fact, it can be shown that if f is a continuous function of two variables and g is a continuous function of a single variable defined on the range of f, then the composite function $h = g \circ f$ defined by $h(x, y) = g(f(x, y))$ is also a continuous function.

EXAMPLE 9 □ Where is the function $h(x, y) = \arctan(y/x)$ continuous?

SOLUTION The function $f(x, y) = y/x$ is a rational function and therefore continuous except on the line $x = 0$. The function $g(t) = \arctan t$ is continuous everywhere. So the composite function

$$g(f(x, y)) = \arctan(y/x) = h(x, y)$$

is continuous except where $x = 0$. The graph in Figure 9 shows the break in the graph of h above the y-axis. □

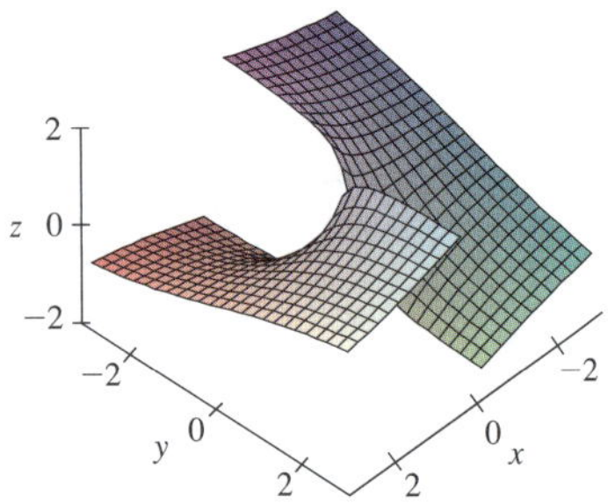

FIGURE 9
The function $h(x, y) = \arctan(y/x)$ is discontinuous where $x = 0$.

Functions of Three or More Variables

Everything that we have done in this section can be extended to functions of three or more variables. The notation

$$\lim_{(x, y, z) \to (a, b, c)} f(x, y, z) = L$$

means that the values of $f(x, y, z)$ approach the number L as the point (x, y, z) approaches the point (a, b, c) along any path in the domain of f. Because the distance between two points (x, y, z) and (a, b, c) in $\mathbb{R}^3$ is given by $\sqrt{(x - a)^2 + (y - b)^2 + (z - c)^2}$, we can write the precise definition as follows: For every number $\varepsilon > 0$ there is a corresponding number $\delta > 0$ such that

$$|f(x, y, z) - L| < \varepsilon \qquad \text{whenever} \qquad 0 < \sqrt{(x - a)^2 + (y - b)^2 + (z - c)^2} < \delta$$

and (x, y, z) is in the domain of f.

The function f is **continuous** at (a, b, c) if

$$\lim_{(x, y, z) \to (a, b, c)} f(x, y, z) = f(a, b, c)$$

For instance, the function

$$f(x, y, z) = \frac{1}{x^2 + y^2 + z^2 - 1}$$

is a rational function of three variables and so is continuous at every point in $\mathbb{R}^3$ except where $x^2 + y^2 + z^2 = 1$. In other words, it is discontinuous on the sphere with center the origin and radius 1.

If we use the vector notation introduced at the end of Section 15.1, then we can write the definitions of a limit for functions of two or three variables in a single compact form as follows.

5 If f is defined on a subset D of $\mathbb{R}^n$, then $\lim_{\mathbf{x}\to\mathbf{a}} f(\mathbf{x}) = L$ means that for every number $\varepsilon > 0$ there is a corresponding number $\delta > 0$ such that

$$|f(\mathbf{x}) - L| < \varepsilon \text{ whenever } \mathbf{x} \in D \text{ and } 0 < |\mathbf{x} - \mathbf{a}| < \delta$$

Notice that if $n = 1$, then $\mathbf{x} = x$ and $\mathbf{a} = a$, and (5) is just the definition of a limit for functions of a single variable. For the case $n = 2$, we have $\mathbf{x} = \langle x, y\rangle$, $\mathbf{a} = \langle a, b\rangle$, and $|\mathbf{x} - \mathbf{a}| = \sqrt{(x-a)^2 + (y-b)^2}$, so (5) becomes Definition 1. If $n = 3$, then $\mathbf{x} = \langle x, y, z\rangle$, $\mathbf{a} = \langle a, b, c\rangle$, and (5) becomes the definition of a limit of a function of three variables. In each case the definition of continuity can be written as

$$\lim_{\mathbf{x}\to\mathbf{a}} f(\mathbf{x}) = f(\mathbf{a})$$

11.2A Exercises

1. Suppose that $\lim_{(x, y)\to(3, 1)} f(x, y) = 6$. What can you say about the value of $f(3, 1)$? What if f is continuous?

2. Explain why each function is continuous or discontinuous.
   (a) The outdoor temperature as a function of longitude, latitude, and time
   (b) Elevation (height above sea level) as a function of longitude, latitude, and time
   (c) The cost of a taxi ride as a function of distance traveled and time

3–4 □ Use a table of numerical values of $f(x, y)$ for (x, y) near the origin to make a conjecture about the value of the limit of $f(x, y)$ as $(x, y) \to (0, 0)$. Then explain why your guess is correct.

3. $f(x, y) = \dfrac{x^2y^3 + x^3y^2 - 5}{2 - xy}$

4. $f(x, y) = \dfrac{2xy}{x^2 + 2y^2}$

5–20 □ Find the limit, if it exists, or show that the limit does not exist.

5. $\lim_{(x, y)\to(5, -2)} (x^5 + 4x^3y - 5xy^2)$

6. $\lim_{(x, y)\to(6, 3)} xy\cos(x - 2y)$

7. $\lim_{(x, y)\to(0, 0)} \dfrac{x^2}{x^2 + y^2}$

8. $\lim_{(x, y)\to(0, 0)} \dfrac{(x + y)^2}{x^2 + y^2}$

9. $\lim_{(x, y)\to(0, 0)} \dfrac{8x^2y^2}{x^4 + y^4}$

10. $\lim_{(x, y)\to(0, 0)} \dfrac{x^3 + xy^2}{x^2 + y^2}$

11. $\lim_{(x, y)\to(0, 0)} \dfrac{xy}{\sqrt{x^2 + y^2}}$

12. $\lim_{(x, y)\to(0, 0)} \dfrac{xy + 1}{x^2 + y^2 + 1}$

13. $\lim_{(x, y)\to(0, 0)} \dfrac{2x^2y}{x^4 + y^2}$

14. $\lim_{(x, y)\to(0, 0)} \dfrac{x^3y^2}{x^2 + y^2}$

15. $\lim_{(x, y)\to(0, 0)} \dfrac{x^2 + y^2}{\sqrt{x^2 + y^2 + 1} - 1}$

16. $\lim_{(x, y)\to(2, 0)} \dfrac{xy - 2y}{x^2 + y^2 - 4x + 4}$

17. $\lim_{(x, y, z)\to(3, 0, 1)} e^{-xy}\sin(\pi z/2)$

18. $\lim_{(x, y, z)\to(0, 0, 0)} \dfrac{x^2 + 2y^2 + 3z^2}{x^2 + y^2 + z^2}$

19. $\lim_{(x, y, z)\to(0, 0, 0)} \dfrac{xy + yz^2 + xz^2}{x^2 + y^2 + z^4}$

20. $\lim_{(x, y, z)\to(0, 0, 0)} \dfrac{xy + yz + zx}{x^2 + y^2 + z^2}$

21–22 □ Use a computer graph of the function to explain why the limit does not exist.

21. $\lim_{(x, y)\to(0, 0)} \dfrac{2x^2 + 3xy + 4y^2}{3x^2 + 5y^2}$

22. $\lim_{(x, y)\to(0, 0)} \dfrac{xy^3}{x^2 + y^6}$

23–24 □ Find $h(x, y) = g(f(x, y))$ and the set on which h is continuous.

23. $g(t) = t^2 + \sqrt{t}$, $\quad f(x, y) = 2x + 3y - 6$

24. $g(t) = \dfrac{\sqrt{t} - 1}{\sqrt{t} + 1}$, $\quad f(x, y) = x^2 - y$

25–26 □ Graph the function and observe where it is discontinuous. Then use the formula to explain what you have observed.

25. $f(x, y) = e^{1/(x-y)}$

26. $f(x, y) = \dfrac{1}{1 - x^2 - y^2}$

27–36 □ Determine the largest set on which the function is continuous.

27. $F(x, y) = \dfrac{1}{x^2 - y}$

28. $F(x, y) = \dfrac{x - y}{1 + x^2 + y^2}$

29. $F(x, y) = \arctan(x + \sqrt{y})$

30. $F(x, y) = \ln(2x + 3y)$

31. $G(x, y) = \sqrt{x + y} - \sqrt{x - y}$

32. $G(x, y) = \sin^{-1}(x^2 + y^2)$

33. $f(x, y, z) = \dfrac{xyz}{x^2 + y^2 - z}$

34. $f(x, y, z) = \sqrt{x + y + z}$

35. $f(x, y) = \begin{cases} \dfrac{x^2y^3}{2x^2 + y^2} & \text{if } (x, y) \neq (0, 0) \\ 1 & \text{if } (x, y) = (0, 0) \end{cases}$

36. $f(x, y) = \begin{cases} \dfrac{xy}{x^2 + xy + y^2} & \text{if } (x, y) \neq (0, 0) \\ 0 & \text{if } (x, y) = (0, 0) \end{cases}$

37–38 □ Use polar coordinates to find the limit. [If (r, θ) are polar coordinates of the point (x, y) with $r \geq 0$, note that $r \to 0^+$ as $(x, y) \to (0, 0)$.]

37. $\displaystyle\lim_{(x, y) \to (0, 0)} \frac{x^3 + y^3}{x^2 + y^2}$

38. $\displaystyle\lim_{(x, y) \to (0, 0)} (x^2 + y^2) \ln(x^2 + y^2)$

39. Use spherical coordinates to find

$$\lim_{(x, y, z) \to (0, 0, 0)} \frac{xyz}{x^2 + y^2 + z^2}$$

40. At the beginning of this section we considered the function

$$f(x, y) = \frac{\sin(x^2 + y^2)}{x^2 + y^2}$$

and guessed that $f(x, y) \to 1$ as $(x, y) \to (0, 0)$ on the basis of numerical evidence. Use polar coordinates to confirm the value of the limit. Then graph the function.

41. Show that the function f given by $f(\mathbf{x}) = |\mathbf{x}|$ is continuous on $\mathbb{R}^n$. [*Hint*: Consider $|\mathbf{x} - \mathbf{a}|^2 = (\mathbf{x} - \mathbf{a}) \cdot (\mathbf{x} - \mathbf{a})$.]

42. If $\mathbf{c} \in V_n$, show that the function f given by $f(\mathbf{x}) = \mathbf{c} \cdot \mathbf{x}$ is continuous on $\mathbb{R}^n$.

Exercises 11.2A □ page 928

1. Nothing; if f is continuous, $f(3, 1) = 6$ **3.** $-\frac{5}{2}$ **5.** 2025
7. Does not exist **9.** Does not exist **11.** 0
13. Does not exist **15.** 2 **17.** 1 **19.** Does not exist
21. The graph shows that the function approaches different numbers along different lines.
23. $h(x, y) = 4x^2 + 9y^2 + 12xy - 24x - 36y + 36 + \sqrt{2x + 3y - 6}$; $\{(x, y) \mid 2x + 3y \geqslant 6\}$
25. Along the line $y = x$ **27.** $\{(x, y) \mid y \neq x^2\}$
29. $\{(x, y) \mid y \geqslant 0\}$ **31.** $\{(x, y) \mid x \geqslant |y|\}$
33. $\{(x, y, z) \mid z \neq x^2 + y^2\}$ **35.** $\{(x, y) \mid (x, y) \neq (0, 0)\}$
37. 0 **39.** 0

11.3 Partial Derivatives

On a hot day, extreme humidity makes us think the temperature is higher than it really is, whereas in very dry air we perceive the temperature to be lower than the thermometer indicates. The National Weather Service has devised the *heat index* (also called the temperature-humidity index, or humidex) to describe the combined effects of temperature and humidity. The heat index I is the perceived air temperature when the actual temperature is T and the relative humidity is H. So I is a function of T and H and we can write $I = f(T, H)$. The following table of values of I is an excerpt from a table compiled by the National Weather Service.

TABLE 1 Heat index I as a function of temperature and humidity

Relative humidity (%)

Actual temperature (°F)	T \ H	50	55	60	65	70	75	80	85	90
	90	96	98	100	103	106	109	112	115	119
	92	100	103	105	108	112	115	119	123	128
	94	104	107	111	114	118	122	127	132	137
	96	109	113	116	121	125	130	135	141	146
	98	114	118	123	127	133	138	144	150	157
	100	119	124	129	135	141	147	154	161	168

If we concentrate on the highlighted column of the table, which corresponds to a relative humidity of $H = 70\%$, we are considering the heat index as a function of the